# Workshop Calculus

## Guided Exploration with Review

### Volume 2

**Springer**
*New York*
*Berlin*
*Heidelberg*
*Barcelona*
*Budapest*
*Hong Kong*
*London*
*Milan*
*Paris*
*Singapore*
*Tokyo*

# Workshop Calculus

## Guided Exploration with Review

### Volume 2

*Nancy Baxter Hastings*
**Dickinson College**

With contributing authors:

**Priscilla Laws**
**Christa Fratto**
**Kevin Callahan**
**Mark Bottorff**

Springer

...oks in Mathematical Sciences

*Series Editors*

Thomas F. Banchoff
*Brown University*

Jerrold Marsden
*California Institute of Technology*

Keith Devlin
*St. Mary's College*

Stan Wagon
*Macalester College*

Gaston Gonnet
*ETH Zentrum, Zürich*

COVER: Cover art by Kelly Alsedek at Dickinson College, Carlisle, Pennsylvania.

Library of Congress Cataloging-in-Publication Data
Baxter Hastings, Nancy.
    Workshop calculus: guided exploration with review/Nancy Baxter
  Hastings.
      p.   cm. — (Textbooks in mathematical sciences)
    Includes bibliographical references and index.
    ISBN 0-387-98349-X (v. 2: softcover: alk. paper)
    1. Calculus.   I. Title.   II. Series.
  QA303.B3598   1998
  515—dc20                      95-47550

Printed on acid-free paper.

Production managed by Francine McNeill; manufacturing supervised by Jacqui Ashri.
Typeset by Matrix Publishing Services, Inc., York, PA.
Printed and bound by Maple-Vail Book Manufacturing Group, Binghamton, NY.
Printed in the United States of America.

9 8 7 6 5 4 3 2 1

ISBN 0-387-98349-X Springer-Verlag New York Berlin Heidelberg  SPIN 10645860

*To my husband,*
*David,*
*and our family,*
*Erica and Mark,*
*Benjamin,*
*Karin and Matthew,*
*Mark, Margie, and Morgan*
*John and Laura.*

# *Preface*

## TO THE INSTRUCTOR

*I hear, I forget.*
*I see, I remember.*
*I do, I understand.*

Anonymous

## OBJECTIVES OF WORKSHOP CALCULUS

1. Impel students to be active learners.

2. Help students to develop confidence about their ability to think about and do mathematics.

3. Encourage students to read, write, and discuss mathematical ideas.

4. Enhance students' understanding of the fundamental concepts underlying the calculus.

5. Prepare students to use calculus in other disciplines.

6. Inspire students to continue their study of mathematics.

7. Provide an environment where students enjoy learning and doing mathematics.

## THE WORKSHOP APPROACH

*Workshop Calculus: Guided Exploration with Review* provides students with a gateway into the study of calculus. The two-volume series integrates a review of basic pre-calculus ideas with the study of concepts traditionally encountered in beginning calculus: functions, limits, derivatives, integrals, and an introduction to integration techniques and differential equations. It seeks to help students develop the confidence, understanding, and skills necessary for using calculus in the natural and social sciences, and for continuing their study of mathematics.

In the workshop environment, students learn by doing and by reflecting on what they have done. In class, no formal distinction is made between classroom and laboratory work. Lectures are replaced by an interactive teaching format, with the following components:

- *Summary discussion:* Typically, the beginning of each class is devoted to summarizing what happened in the last class, reviewing important ideas, and presenting additional theoretical material. Although this segment of a class may take only 10 minutes or so, many students claim that it is one of the most important parts of the course, since it helps them make connections and focus on the overall picture. Students understand, and consequently value, the discussion because it relates directly to the work that they have been doing.

- *Introductory remarks:* The summary discussion leads into a brief introduction to what's next. The purpose of this initial presentation is to help guide students' thoughts in appropriate directions without giving anything away. New ideas and concepts are introduced in an intuitive way, without giving any formal definitions, proofs of theorems, or detailed examples.

- *Collaborative activities:* The major portion of the class is devoted to students working collaboratively on the tasks and exercises in their *Workshop Calculus* book. As students work together, the instructor moves from group to group, guiding discussions, posing questions, and responding to queries.

Workshop Calculus is part of Dickinson College's Workshop Mathematics Program, which also includes Workshop Statistics and Workshop Quantitative Reasoning, developed by my colleague Allan Rossman. Based on our experiences and those of others who have taught workshop courses, we have developed the following helpful list, which Allan calls "A Dozen (Plus or Minus Two) Suggestions for Workshop Instructors":

- *Take control of the course.* Perhaps this goes without saying, but it is very important for an instructor to establish that he or she has control of the course. It is a mistake to think of Workshop courses as self-paced where the instructor plays but a minor role.

- **Keep the class roughly together.**  We suggest that you take control of the course in part by keeping the students roughly together with the material, not letting some groups get too far ahead or lag behind.

- **Allow students to discover.**  We encourage you to resist the temptation to tell students too much. Rather, let them do the work to discover ideas for themselves. Try not to fall into let-me-show-you-how-to-do-this mode.

- **Promote collaborative learning among students.**  We suggest that you have students work together on the tasks in pairs or groups of three. We do recommend, however, that students be required to write their responses in their books individually.

- **Encourage students' guessing and development of intuition.**  We believe that much can be gained by asking students to think and make predictions about issues before analyzing them in detail.

- **Lecture when appropriate.**  By no means do we propose never speaking to the class as a whole. As a general rule, however, we advocate lecturing on an idea only after students have begun to grapple with it first themselves.

- **Have students do some work by hand.**  While we strongly believe in using technology to explore mathematical phenomena, we think students have much to gain by first becoming competent at performing computations, doing symbolic manipulations, and sketching graphs by hand.

- **Use technology as a tool.**  The counterbalance to the previous suggestion is that students should come to regard technology as an invaluable tool for analyzing functions.

- **Be proactive in approaching students.**  As your students work through the tasks, we strongly suggest that you mingle with them. Ask questions. Join their discussions.

- **Give students access to "right" answers.**  Some students are fearful of a self-discovery approach because they worry about discovering "wrong" things. We appreciate this objection, for it makes a strong case for providing students with regular and consistent feedback.

- **Provide plenty of feedback.**  An instructor can supply much more personalized, in-class feedback with the workshop approach than in a traditional lecture classroom, and the instructor is positioned to continually assess how students are doing. We also encourage you to collect a regular sampling of tasks and homework exercises as another type of feedback.

- **Stress good writing.**  We regard writing-to-learn as an important aspect of a workshop course. Many activities call for students to write interpretations and to explain their findings. We insist that students relate these to the context at hand.

- *Implore students to read well.* Students can do themselves a great service by taking their time and reading not only the individual questions carefully, but also the short blurbs between tasks, which summarize what they have done and point the way to what is to come.

- *Have fun!* We enjoy teaching more with the workshop approach than with lecturing, principally because we get to know the students better and we love to see them actively engaged with the material. We genuinely enjoy talking with individual students and small groups of students on a regular basis, as we try to visit each group several times during a class period. We sincerely hope that you and your students will have as much fun in a Workshop Mathematics course.

## INSTRUCTIONAL MATERIALS

The *Workshop Calculus* book, which we refer to as an "activity guide," is a collection of guided inquiry notes presented in a workbook format. As students begin to use the book, encourage them to tear out the pages for the current section and to place them in a three-ring binder. These pages can then be interspersed with lecture and discussion notes, responses to homework exercises, supplemental activities, and so on. During the course, they will put together their own books.

Each section in the book consists of a sequence of tasks followed by a set of homework exercises. These activities are designed to help students think like mathematicians—to make observations and connections, ask questions, explore, guess, learn from their errors, share ideas, read, write, and discuss mathematics.

The tasks are designed to help students explore new concepts or discover ways to solve problems. The steps in the tasks provide students with a substantial amount of guidance. Students make predictions, do calculations, and enter observations directly in their guide. At the conclusion of each task, the main ideas are summarized, and students are given a brief overview of what they will be doing in the next task. The tasks are intended to be completed in a linear fashion.

The homework exercises provide students with an opportunity to utilize new techniques, to think more deeply about the concepts introduced in the section, and occasionally to tackle a new idea. New information that is presented in an exercise is usually not needed in subsequent tasks. However, if a subsequent task does rely on a concept introduced in an exercise, students are referred back to the exercise for review. Any subset of the exercises may be assigned, and they may be completed in any order or interspersed with associated tasks. The homework exercises should probably be called "post-task activities," since the term "homework" implies that they are to be done outside of class. This is not our intention; both tasks and exercises may be completed either in or out of class.

As the conclusion to each unit, students reflect on what they have learned in their "journal entry" for the unit. They are asked to describe in their own words the concepts they have studied, how they fit together, which ones were easy, and which were hard. They are also asked to reflect on the learning environment for the course. We view this activity as one of the most important in a unit. Not only do the journal entries provide us with feedback and enable us to catch any last misconceptions, but more importantly, they provide the students with an opportunity to think about what has been going on and to write about their observations.

Workshop Calculus utilizes several software tools. In Unit 1, students use a motion detector to create distance versus time functions and to analyze their behavior. Throughout the course, they use a computer algebra system (CAS) to do symbolic, numerical, and graphical manipulations, and they use the mathematical programming language ISETL to construct functions and to form mental images associated with abstract mathematical ideas, such as the process of a function or the limiting behavior of a function. Although the Workshop Calculus materials use particular software tools, there are no references in the text to a particular type of computer platform or to a specific CAS package.

Ideally, a computer should be available for each group of two to four students. Each machine should be equipped with a Microcomputer-Based Laboratory (MBL) interface and software to support ultrasonic motion detection;[1] a computer algebra system, such as *Mathematica*®, Maple, or Derive; and the programming language ISETL.[2]

Model handouts for each of the software tools, including "Using the Motion Detector in Workshop Calculus," "Using ISETL in Workshop Calculus," and "Using *Mathematica* in Workshop Calculus," are available electronically from our Web site at Dickinson College. Each handout provides an overview of the features of the software tool that are used in Workshop Calculus. Although these handouts were developed for use at Dickinson College, they can be easily customized for use at other institutions. Most of the information contained in the handouts is *not* contained in the activity guide, since it depends on the particular system and tools being used. Students, however, will need this information to do the tasks. You will need to download the Dickinson College version of the handouts, make the necessary modifications for your locale, and distribute copies to your students.

---

[1] Available for both Macintosh and PC platforms from Vernier Software, 8565 S.W. Beaverton-Hillsdale Highway, Portland, OR 97225.

[2] ISETL is freeware. Mac and DOS versions are available from http://www.math.purdue.edu/~ccc/distribution.html. The Windows® version is available from http://csis03.muc.edu/isetlw/isetlw.htm.

In addition to the handouts on the various software tools, a set of notes to the instructor is also available electronically. These notes contain topics for discussion and review; suggested timing for each task; solutions to homework exercises; and sample schedules, syllabi, and exams.

## ACKNOWLEDGMENTS

The Workshop Calculus materials were developed in consultation with my physics colleague, Priscilla Laws, and one of my former students, Christa Fratto. Priscilla was the impelling force behind the project. She developed many of the applications that appear in the text, and her award-winning Workshop Physics project provided a model for the Workshop Calculus materials and the underlying pedagogical approach. She and Ron Thornton of Tufts University modified the MBL tools—which they had developed for use by physics students—for use in Workshop Calculus.

Christa Fratto, who graduated from Dickinson College in 1994, started working on the materials as a Dana Student Intern, during the summer of 1992. She quickly became a partner in the project. She tested activities, offered in-depth editorial comments, developed problem sets, helped collect and analyze assessment data, and supervised the student assistants for the Workshop Calculus classes. Following graduation, she has continued to work on the project, writing the handouts for the software tools and the answer key for the homework exercises, and developing new tasks.

Other major contributors include Kevin Callahan and Mark Bottorff, who helped design, write, and test initial versions of the material while on the faculty at Dickinson College. Kevin is now using the materials at California State University at Hayward, and Mark is pursuing his Ph.D. degree in Mathematical Physics.

A number of other colleagues have tested the materials and provided constructive feedback. Those who tested the materials and offered helpful comments include Peter Martin, Shari Prevost, Judy Roskowski, Barry Tesman, Jack Stodghill, and Blayne Carroll at Dickinson College; Carol Harrison at Susquehanna University; Nancy Johnson at Lake Brantley High School; Michael Kantor at Knox College; Stacy Landry at The Potomac School; Sandy Skidmore and Julia Clark at Emory and Henry College; Sue Suran at Gettysburg High School; Sam Tumolo at Cincinnati Country Day School; and Barbara Wahl at Hanover College. The development of the materials was also influenced by helpful suggestions from Ed Dubinsky of Purdue University, who served as the project's mathematics education research consultant, and David Smith of Duke University, who served as the project's outside evaluator.

The Dickinson College students who assisted in Workshop Calculus classes helped make the materials more learner-centered and more user-friendly. These students include Jennifer Becker, Jason Cutshall, Amy Demski, Kimberly Kendall, Greta Kramer, Russell LaMantia, Tamara Manahan, Susan Nouse, Alexandria Pefkaros, Benjamin Seward, Melissa Tan, Katharyn Wilber, and Jennifer Wysocki. In addition, Kathy Clawson, Hannah Hazard, Jennifer Hoenstine, Linda Mellott, Marlo Mewherter, Matthew Parks, and Katherine Reynolds worked on the project as Dana Student Interns, reviewing the materials, analyzing assessment data, developing answer keys, and designing Web pages. Virginia Laws did the initial version of the illustrations, and Matthew Weber helped proofread the final copy.

An important aspect of the development of the Workshop Calculus materials is the ongoing assessment activities. With the help of Jack Bookman, who served as the project's outside evaluation expert, we have analyzed student attitudes and learning gains, observed gender differences, collected retention data, and examined performance in subsequent classes. The information has provided the program with documented credibility and has been used to refine the materials for publication.

The Workshop Mathematics Program has received generous support from the U.S. Department of Education's Fund for Improvement of Post Secondary Education (FIPSE #P116B50675 and FIPSE #P116B11132), the National Science Foundation (NSF/USE #9152325, NSF/DUE #9450746, and NSF/DUE #9554684), and the Knight Foundation. For the past five years, Joanne Weissman has served as the project manager for the Workshop Mathematics Program. She has done a superb job, keeping the program running smoothly and keeping us focused and on task.

Publication of Volume 2 of the Workshop Calculus activity guide marks the culmination of seven years of testing and development. We have enjoyed working with Jerry Lyons, Editorial Director of Physical Sciences at Springer-Verlag. Jerry is a kindred spirit, who shares our excitement and understands our vision. We appreciate his support, value his advice, and enjoy his friendship. And, finally, we wish to thank Kim Banister, from Dickinson College, who did the illustrations for the manuscript. In her drawings, she caught the essence of the workshop approach: students exploring mathematical ideas, working together, and enjoying the learning experience.

Nancy Baxter Hastings
Professor of Mathematics and
Computer Science
baxter@dickinson.edu

# *Preface*

# TO THE STUDENT

*Everyone knows that if you want to do physics or engineering, you had better be good at mathematics. More and more people are finding out that if you want to work in certain areas of economics or biology, you had better brush up on your mathematics. Mathematics has penetrated sociology, psychology, medicine and linguistics . . . it has been infiltrating the field of history. Why is this so? What gives mathematics its power? What makes it work?*

*. . . the universe expresses itself naturally in the language of mathematics. The force of gravity diminishes as the second power of the distance; the planets go around the sun in ellipses, light travels in a straight line. . . . Mathematics in this view, has evolved precisely as a symbolic counterpart of this universe. It is no wonder then, that mathematics works: that is exactly its reason for existence. The universe has imposed mathematics upon humanity. . . .*

Philip J. Davis and Rubin Hersh
Co-authors of *The Mathematical Experience*
Birkhauser, Boston, 1981

## Why Study Calculus?

Why should you study calculus? When students like yourself are asked their reasons for taking calculus courses, they often give reasons such as, "It's required for my major." "My parents want me to take it." "I like math." Mathematics teachers would love to have more students give idealistic answers such as, "Calculus is a great intellectual achievement that has made major contributions to the development of philosophy and science. Without an understanding of calculus and an appreciation of its inherent beauty, one cannot be considered an educated person."

Although most mathematicians and scientists believe that becoming an educated person ought to be the major reason why you should study calculus, we can think of two other equally important reasons for studying this branch of mathematics: (1) mastering calculus can provide you with conceptual tools that will contribute to your understanding of phenomena in many other fields of study, and (2) the process of learning calculus can help you acquire invaluable critical thinking skills that will enrich the rest of your life.

## What Is Calculus?

Basically, calculus is a branch of mathematics that has been developed to describe relationships between things that can change continuously. For example, consider the mathematical relationship between the diameter of a pizza and its area. You know from geometry that the area of a perfectly round pizza is related to its diameter by the equation $A = \frac{1}{4}\pi d^2$. You also know that the diameter can be changed continuously. Thus, you don't have to make just 9" pizzas or 12" pizzas. You could decide to make one that is

10.12" or one that is 10.13", or one whose diameter is halfway between these two sizes. A pizza maker could use calculus to figure out how the area of a pizza changes when the diameter changes, a little more easily than a person who only knows geometry.

But it is not only pizza makers who might benefit by studying calculus. Someone working for the Federal Reserve might want to figure out how much metal would be saved if the size of a coin is reduced. A biologist might want to study how the growth rate of a bacterial colony in a circular petri dish changes over time. An astronomer might be curious about the accretion of material in Saturn's famous rings. All of these questions can be answered by using calculus to find the relationship between the change in the diameter of a circle and its area.

## What Are You Expected to Know?

As you begin this volume, we expect you to have a firm understanding of what a function is and what a limit is (see Volume 1, Units 1 through 4).

Calculus is a study of functions. You should be able to determine if a given relationship defines a function, and if it does, you should be able to identify its domain and range. You should be familiar with various ways of representing a function—symbolically, graphically, and verbally—and you should be familiar with various ways of constructing a function—using an expression, piecewise-definition, graph, table, and set of ordered pairs. You should be familiar with the various classes of functions—polynomial functions, power functions, rational functions, trigonometric functions, exponential functions, and logarithmic functions—and you should know how to create a new function by combining old ones and by reflecting a function through a line. You should be able to use a function to model a situation.

In addition, you should be familiar with the terminology used for analyzing the properties of a function. Given the graph of a function, you should be able to identify the intervals where a function is continuous, where it is increasing and decreasing, and where it is concave up and concave down, and you should be able to locate a function's points of discontinuity, its local extrema, and its inflection points. Moreover, you should have a conceptual understanding of the notion of a tangent line to a graph and be able to relate the behavior of the tangent line as it travels along the graph to the properties of the function—for instance, if the slope of the tangent line is positive, then the function is increasing. You should be able to identify situations where the tangent line does not exist—for instance, the tangent line does not exist at a cusp.

Limits, on the other hand, provide the bridge from algebra and geometry to calculus. You will use the concept of limit throughout your study of calculus. At this point, you should be able to find a reasonable value for a limit using an input/output table and using a graphic approach. You should be able to identify situations where a limit does not exist—for instance, at a "jump" or a vertical asymptote—and explain what happens in these cases in terms of the left- and right-hand limits. You should understand the limit-based definition for continuity and be able to describe the limiting behavior of a function at a place where it is not continuous—that is, where the function has a removable, jump, or blowup discontinuity. You should be able to evaluate the limit of a continuous function using substitution and the limit of a rational function near a hole in the graph. You should be able to use limits to analyze the shape of the graph of a function near a vertical asymptote and to locate horizontal asymptotes.

Unit 5 provides a first look at the two major calculus concepts—derivatives and definite integrals—and how they are connected. It is the swing unit in the Workshop Calculus series, as it is the last unit in Volume 1 and the first in Volume 2.

As you complete the activities in this volume, you will learn a lot about the nature of calculus. You will discover rules for finding derivatives and evaluating integrals, and you will investigate how to use calculus to solve problems.

## Using Computers and Collaboration to Study Calculus

The methods used to teach Workshop Calculus may be new to you. In the workshop environment, formal lectures will be replaced by an interactive teaching format. You will learn by doing and by reflecting on what you have done. Initially, new ideas will be introduced to you in an informal and intuitive way. You will then work collaboratively with your classmates on the activities in this workbook, exploring and discovering mathematical concepts on your own. You will be encouraged to share your observations during class discussions.

Although we can take responsibility for designing a good learning environment and for attempting to teach you calculus, you must take responsibility for learning it. No one else can learn it for you. You should find the thinking skills and mathematical techniques acquired in this course useful in the future. Most importantly, we hope you enjoy the study of calculus and begin to appreciate its inherent beauty.

A number of the activities in this course will involve using computers to enhance your learning. These tools include the mathematical programming language ISETL and a computer algebra system. Using the computer will help you develop a conceptual understanding of important mathematical concepts and help you focus on significant ideas, rather than spending a lot of time on extraneous details.

## Some Important Advice Before You Begin

- *Put together your own book.* Remove the pages for the current section from your activity guide, and place them in a three-ring binder. Intersperse the pages with lecture and discussion notes, answers to homework problems, and handouts from your instructor.

- *Read carefully the short blurbs at the beginning of each section and prior to each task.* These blurbs summarize what you have done and point the way to what is to come. They contain important and useful information.

- *Work closely with the members of your group.* Think about the tasks together. Discuss how you might respond to a given question. Share your thoughts and your ideas. Help one another. Talk mathematics.

- *Answer the questions in your activity guide in your own words.* Work together, but when it comes time to write down the answer to a question, do not simply copy what one of your partners has written.

- *Use separate sheets of paper for homework problems.* Unless otherwise instructed, do not try to squish the answers in between the lines in your activity guide.

- *Think about what the computer is doing.* Whenever you ask the computer to perform a task, think about how the computer might be processing the information that you have given it, keeping in mind:

  —What you have commanded the computer to do.
  —Why you asked it to do whatever it is doing.
  —How it might be doing whatever you have told it to do.
  —What the results mean.

- ***Switch typists on a regular basis.***   Do not become the designated typist or an ongoing observer for your group. Learning to use the computer is an important part of the learning process.

- ***Have fun!***

Nancy Baxter Hastings
Professor of Mathematics and
Computer Science
baxter@dickinson.edu

# Contents

# Volume
# 2

# Unit 5:

# DERIVATIVES AND INTEGRALS: FIRST PASS

*Every natural phenomenon, from quantum vibrations of subatomic particles to the universe itself, is a manifestation of change. Developing organisms change as they grow. Populations of living creatures, from viruses to whales, vary from day to day or from year to year. Our history, politics, economics, and climate are subject to constant, and often baffling, changes. Some changes are simple: the cycles of the season, the ebb and flow of the tides. Others seem more complicated: economic recessions, outbreaks of diseases, the weather. All kinds of changes influence our lives. ... The traditional approach to the mathematics of change can be summed up in one word: calculus.*

Ian Stewart, "Change." In Lynn Steen (Ed.): *On the Shoulders of Giants: New Approaches to Numeracy*, National Academy Press, 1990, 183–184.

## OBJECTIVES

1. Develop a definition for the derivative.

2. Analyze the behavior of the derivative. Find its domain. Represent it by a graph and an expression.

3. Examine the relationship between a function and its derivatives.

4. Develop a definition for the definite integral.

5. Examine the relationship between a function and its associated accumulation function.

6. Consider the connection between the derivative and the definite integral.

## OVERVIEW

This unit provides a first pass at the two basic concepts underlying the study of calculus—the derivative and the definite integral—and the connection between these two ideas.

Actually, you already know a lot about derivatives, under the guise of rates of change and slopes of tangent lines. The problem is that you do not know how to find the actual value of a derivative. For example, you cannot use the two-point formula for slope (which you learned in algebra) to find the slope of the tangent line. Why? Because in the case of a tangent line, you only know one point, namely the point where the tangent line touches the graph. You can use limits, however, to get around this problem.

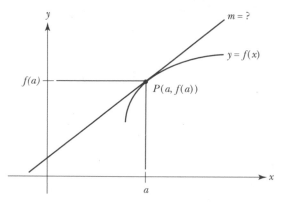

Figure 1.

The idea is to construct a sequence of secant lines that get closer and closer the tangent line. You can calculate the slopes of these lines, since they intersect the graph at two points. You can then find the value of the slope of the tangent line by taking the limit of the slopes of the secant lines as the secant lines approach the tangent line.

The second major concept you will study in this unit is the definite integral. Whereas derivatives are associated with rates of change, integrals correspond to the accumulation of quantities. Integrals can be used, for example, to find the area of an oddly shaped region, such as the shaded region in Figure 2. You know formulas for finding the area of regular shapes, such as rectangles, triangles, and circles, but there isn't a formula for finding the area of the region that lies under the graph of $y = f(x)$ and over the closed interval $[a,b]$. You can, however, approximate this area by covering it with shapes, whose areas you can calculate, such as rectangles.

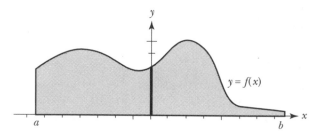

Figure 2.

You can then find the exact value for the area by making the approximations become more and more precise as you take the limit. This is the same approach you will use to find the slope of the tangent line: First approximate the value you want by applying a formula you already know, and then find the exact value by making the approximations become more exact as you take the limit. You will use this approach again and again throughout your study of calculus.

The objectives of this unit are to develop a conceptual understanding of the fundamental concepts, derivative and definite integral, and to make a first pass at examining how these two important ideas are related. Understanding what these ideas mean will help you recognize where they apply. Exploring ways to evaluate them directly and how to use them in a variety of situations is the focus of subsequent units.

## SECTION 1

### The Derivative

In Unit 1 you developed a conceptual understanding of the notion of the tangent line to a curve at a point $P$. You observed that if you repeatedly zoom in and magnify the portion of the curve containing $P$, the curve "straightens out" and merges with the tangent line to the curve at $P$. You noted that the tangent line does not exist at a point where the function has a sharp peak or dip—that is, it does not exist at a *cusp*—because no matter how close you zoom in to a point of this type, the cusp is always there and the graph of the curve never straightens. In HW1.13 and HW4.5, you approximated the value of the slope of the tangent line at a given point by finding the slope of the secant line determined by the given point and a point nearby.

What does all this have to do with derivatives? The value of the derivative of a function at a given point represents the value of the slope of the tangent line to the curve at that point. It also gives the rate of change of the function at the point. In other words,

derivative of $f$ at $x$ equals $a$
    = slope of the tangent line to the graph of $f$ at $P(a, f(a))$
    = rate of change of $f$ at $x$ equals $a$

The question, as you know, is how can you find the exact value? In the next task, you will use limits to find the slope of the tangent line for a specific function. In the following task, you will develop a definition for derivative by formalizing the approach that you develop in Task 5-1. Note: Before beginning the next task, review the ideas developed in homework problems HW1.13 and HW4.5.

### Task 5-1: Examining an Example

Consider $f(x) = x^2 + 3$ at $x = 1$. Use a graphic and a numeric approach to examine the limiting behavior of the slopes of the secant lines determined by $P(1,4)$. Based on your observations, find a reasonable value for the slope of the tangent line at $x = 1$. Use a symbolic approach to find the actual value of the slope. Show that the three approaches—graphic, numeric, and symbolic—yield the same result.

1. Use a graphic approach to examine the limiting behavior of the slopes of the secant lines through $P(1,4)$.

   **a.** Consider the graph of $f$ given below. The sequence of points indicated on the horizontal axis approach 1 from the left and from the right. Mark the points on the graph that correspond to the items in the sequence. Use a pencil and straightedge to draw the secant lines determined by $P(1,4)$ and each of these points.

   *Note: Secant lines are not line segments. They go on indefinitely. Extend each line through the two points which determine it.*

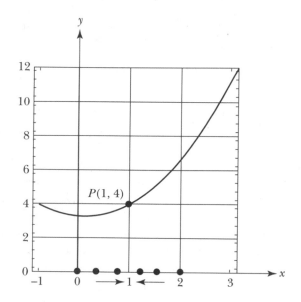

   **b.** Based on the behavior of the secant lines, find a reasonable value for the slope of the tangent line to the graph of $f(x) = x^2 + 3$ at $x = 1$. *Caution*: The two axes have different scales.

2. Try this again, only this time use a numeric approach to examine the limiting behavior of the slopes of the secant lines through $P(1,4)$.

   **a.** Run the following ISETL session. Think carefully about what is going on here.

| ISETL<br>Code | Predicted<br>Output | Actual<br>Output |
|---|---|---|
| $ Define the function $f(x) = x^2 + 3$.<br><br>`f := func (x);`<br>`    return x**2 + 3;`<br>`end func;`<br><br>$ Set $a = 1$.<br><br>`a := 1;` | | |
| $ Define two sequences of values, one approaching 1 from the left and one from the right.<br><br>`n := 5;`<br><br>`LeftSeq := [a – 10**(–i) : i in [1..n]];`<br>`LeftSeq;`<br><br>`RightSeq := [a + 10**(–i) : i in [1..n]];`<br>`RightSeq;` | | |
| $ Create two tables, one corresponding to LeftSeq and one to RightSeq. Loop through the<br>$ items in each sequence. For each value of x, calculate the slope of the secant line<br>$ determined by $P(a,f(a))$ and $Q(x,f(x))$.<br><br>`for x in LeftSeq do`<br>`    writeln x, (f(x) – f(a))/(x – a);`<br>`end for;`<br><br>`for x in RightSeq do`<br>`    writeln x, (f(x) – f(a))/(x – a);`<br>`end for;` | | |

**b.** Print copies of the two tables. Place them in your Activity Guide. Label the columns in the tables.

**c.** Based on your tables, draw some conclusions about the limiting behavior of the slopes of the secant lines determined by $P(a, f(a))$ and $Q(x, f(x))$ when $a = 1$.

Slope of secant lines
determined by $P(1,4)$ $\rightarrow$ ☐ as $x \rightarrow 1^-$

and

$$\begin{array}{l}\text{Slope of secant lines} \\ \text{determined by } P(1,4)\end{array} \rightarrow \boxed{\phantom{xxx}} \quad \text{as } x \rightarrow 1^+$$

Therefore,

$$\begin{array}{l}\text{Slope of secant lines} \\ \text{determined by } P(1,4)\end{array} \rightarrow \boxed{\phantom{xxx}} \quad \text{as } x \rightarrow 1$$

**d.** Does the conclusion you reached using a numeric approach support the conclusion you reached in part 1 using a graphic approach? If not, look over your work again.

3. One more time. Use a symbolic approach to find the exact value of the slope of the tangent line to the $f(x) = x^2 + 3$ at $x = 1$.

**a.** Find a general expression for the slope of the secant line through $P(1,4)$ and $Q(x, f(x))$, where $f(x) = x^2 + 3$; that is, find an expression (in terms of $x$) for

$$\frac{f(x) - f(1)}{x - 1}$$

**b.** Calculate the actual value of the slope of the tangent line to the graph of $f(x) = x^2 + 3$ at $x = 1$ by finding the limit of the general expression for the slope of the secant lines (which you developed above in part a) as $x \rightarrow 1$.

**c.** This result should agree with the conjectures you made based on the graph in part 1 and your ISETL tables in part 2. Does it? If not, try to find where you went astray....

In the specific example considered in the last task, you analyzed the behavior of the slopes of the secant lines as they approached the tangent line to the graph at a particular point. You estimated the slope of the tangent line using graphic and numeric approaches, and then found the exact value by taking the appropriate limit. In the next task, you will generalize your observations and, in the process, discover a definition for the derivative.

The underlying idea is as follows: As long as the graph of $f$ is smooth at $x = a$—that is, as long as $f$ is continuous and its graph does not have a sharp peak or dip at $x = a$—you can:

- Construct two sequences, one approaching $a$ from the left and one from the right.
- Find an expression for the slope of the secant line determined by $P(a, f(a))$ and the point on the graph corresponding to an item in the sequence.
- Find the slope of the tangent line by taking the limit of the slopes of the secant lines as the items in the sequence move closer and closer to $a$.

This is the same approach that you used in the last task to find the slope of the tangent line to the graph of $f(x) = x^2 + 3$ at $x = 1$. There will be one notational change, however. The items in the sequence that get closer and closer to $a$ will be defined in terms of $a + h$, where $h \rightarrow 0$, instead of in terms of $x$ where $x \rightarrow a$. This change in notation reflects the way LeftSeq and RightSeq are constructed.

## Task 5-2: Discovering a Definition for the Derivative

**1.** Construct a general sequence that approaches $x = a$ from both sides.

**a.** Consider the sequence

   RightSeq:= [a + 10**(−i): i in [1..5]];

   The items in RightSeq approach $x = a$ from the right. Each item in RightSeq is formed by adding a small positive number $h$ to $a$, that is, each item has the form

$$a + h, \quad \text{where in this case } h = 10^{-i}$$

Analyze the limiting behavior of $a + h$ as $h$ approaches 0 from the right.

   **(1)** Let $h$ be a small positive number. Mark $a + h$ on the axis below.

   **(2)** Describe what happens to the value of $a + h$ as $h \to 0^{+}$. Support your description by drawing an arrow on the diagram in part (1).

**b.** Now look to the left of $a$. A sequence such as

   LeftSeq := [a − 10**(−i): i in [1..5]];

approaches $x = a$ from the left, where the items in LeftSeq have the form

$$a + h, \quad \text{where } h = -(10^{-i})$$

   Each item in the sequence is formed by adding to $a$ negative number $h$ close to 0. Analyze the limiting behavior of $a + h$ as $h$ approaches 0 from the left.

   **(1)** Let $h$ be a negative number close to 0. Mark $a + h$ on the axis below.

   **(2)** Describe what happens to the value of $a + h$, as $h \to 0^{-}$. Support your description by drawing an arrow on the diagram in part (1).

**c.** Describe, in general, the limiting behavior of $a + h$ as $h \to 0$—that is, as $h$ approaches 0 from both sides.

2. Find a general expression for the slope of the secant line determined by $x = a$ and a point close to $a$.

a. Consider $a + h$, which is near $a$ but to the right of $a$.

(1) Sketch the secant line corresponding to $a$ and $a + h$. Label the diagram given below as follows:

(a) Mark $a + h$ on the x-axis, where $h$ is a small positive number.

(b) Mark $f(a)$ and $f(a + h)$ on the y-axis.

(c) Label the points $P(a, f(a))$ and $Q(a + h, f(a + h))$ on the graph of the function.

(d) Sketch the secant line determined by $P$ and $Q$.

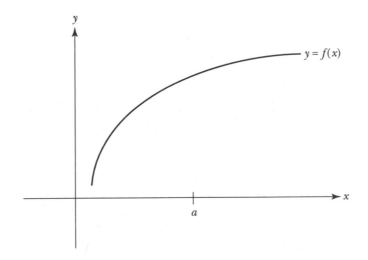

(2) Find an expression for the slope of the secant line in terms of the coordinates of $P$ and $Q$—that is, in terms of $a$, $f(a)$, $a + h$, and $f(a + h)$. Simplify your expression.

(3) Describe the relationship between the values of the slopes of the secant lines as $h \to 0^+$ and the value of the slope of the tangent line at the point $P(a, f(a))$.

b. Do the same thing that you did above, but this time approach $a$ from the left; that is, consider $a + h$, which is near $a$ but to the left of $a$.

(1) Sketch the secant line corresponding to $a$ and $a + h$. Label the diagram given below as follows:

(a) Mark $a + h$ on the x-axis, where $h$ is a negative number close to 0.

(b) Mark $f(a)$ and $f(a + h)$ on the y-axis.

(c) Label the points $P(a, f(a))$ and $R(a + h, f(a + h))$ on the graph of the function.

(d) Sketch the secant line determined by $P$ and $R$.

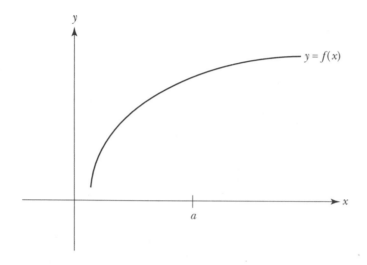

(2) Find the slope of the secant line in terms of the coordinates of $P$ and $R$. Simplify your expression.

(3) Describe the relationship between the values of the slopes of the secant lines as $h \to 0^-$ and the value of the slope of the tangent line at the point $P(a, f(a))$.

**3.** Find the slope of the tangent line by taking the appropriate limit of the expression for the slope of the secant line determined by $x = a$ and $x = a + h$.

Congratulations! You have just developed the notation for what mathematicians call the *derivative*, one of the most important and fundamental concepts in calculus.

In particular, the derivative of a function $y = f(x)$ at $x = a$ is given by

$$\lim_{h \to 0} \underbrace{\frac{f(a + h) - f(a)}{h}}_{\substack{\text{slope of secant line deter-} \\ \text{mined by } P(a, f(a)) \text{ and} \\ Q(a + h, f(a + h))}}$$

slope of tangent line to graph of $f$ at $P(a, f(a))$

provided the limit exists. If the limit exists, $f$ is said to be *differentiable* at $x = a$ and the limit can be denoted by $f'(a)$, which is read "$f$ prime at $a$" or "the derivative of $f$ at $a$." Evaluating a limit of this form gives a number, such as 10 or $-1.5$ or 23.6. Although there has been a lot of talk about tangent lines, keep in mind that there are several ways to interpret this number. It equals the slope of the tangent line at $x = a$. It represents the rate of change of the function at $x = a$. And, more generally, it is the value of the derivative at $x = a$.

The derivative takes an $x$-value in the domain of $f$ and, if the limit exists, returns the value of the slope of the tangent line to the graph of $f$ at that point or the value of the rate of change at $f$ at that point. Consequently, the *derivative of a function is itself a function*, where the domain of the derivative is contained in the domain of the underlying function. If $f$ is the name of the underlying function, then the derivative of $f$ is denoted by $f'$, and you can find the *general derivative* of $f$, $f'(x)$, by evaluating the limit

$$f'(x) = \lim_{h \to 0} \frac{f(x + h) - f(x)}{h}$$

If $f$ is a polynomial, evaluating a limit of this form leads to an expression, such as $2x$ or $x^2 + 6$. You can use the expression representing $f'$ to find the value of the derivative at a particular input, such as $x = 2$.

In general, to find $f'(x)$ using the definition of derivative, you do the following:

• Write down the *difference quotient*

$$\frac{f(x + h) - f(x)}{h}$$

• Simplify the difference quotient. Try to eliminate the $h$ in the denominator so you can evaluate the limit by substituting zero for $h$.
• Take the limit as $h \to 0$.

In the next task, you will represent the derivatives of some functions by expressions and then use the expressions to analyze the functions.

## Task 5-3: Representing the Derivative by Expressions

Find the derivative of a function using the definition. Use the derivative to calculate some tangent lines and analyze the behavior of the function. Before you begin, recall how to evaluate $f(x + h)$.

1. If $f(x) = -2x^2 + x - \frac{1}{2}$, you find $f(x + h)$ by evaluating $f$ at $x + h$; that is, by substituting $x + h$ in for $x$.

$$f(x + h) = -2(x + h)^2 + (x + h) - \tfrac{1}{2}$$
$$= -2(x^2 + 2xh + h^2) + (x + h) - \tfrac{1}{2}$$
$$f(x + h) = -2x^2 - 4xh - 2h^2 + x + h - \tfrac{1}{2}$$

Evaluate each of the following functions at $x + h$. Simplify the result, if possible.

**5**

**a.** $f(x) = 5x - 1$

**b.** $f(x) = 2 - 4x - x^2$

**c.** $f(x) = -99$

**d.** $f(x) = \sqrt{x + 4}$

**e.** $f(x) = \frac{1}{2}\sin(2x)$

**f.** $f(x) = 2^x$

**g.** $f(x) = \dfrac{x^2 - 1}{3x + 6}$

2. Consider $f(x) = x^2 + 3$, which you considered in Task 5-1.

   **a.** Use the definition of derivative to find an expression for $f'(x)$.

$$f'(x) = \lim_{h \to 0} \frac{f(x + h) - f(x)}{h} \qquad \text{Definition of } f'(x)$$

$$= \lim_{h \to 0} \rule{4cm}{0.4pt} \qquad \begin{array}{l}\text{Evaluate } f(x + h) \\ \text{and } f(x)\end{array}$$

$$= \lim_{h \to 0} \rule{4cm}{0.4pt} \qquad \text{Expand terms}$$

$$= \lim_{h \to 0} \rule{4cm}{0.4pt} \qquad \begin{array}{l}\text{Simplify, factoring} \\ h \text{ out of the numerator}\end{array}$$

$$= \lim_{h \to 0} \rule{2cm}{0.4pt} \qquad \text{Cancel } h\text{'s}$$

$$f'(x) = \rule{2cm}{0.4pt} \qquad \text{Take limit}$$

   **b.** Use the derivative to find the tangent line to the graph of $f$ at the point $P(1,4)$.

      **(1)** Find the slope of the tangent line to the graph of $f$ when $x = 1$; that is, find $f'(1)$.

(2) Use the point-slope formula to find the equation of the tangent line to the graph of $f$ when $x = 1$.

*Hint: The tangent line touches the graph of f at P(1,4). You found its slope in part (1).*

(3) Use your CAS to graph $f$ and the tangent line to the graph of $f$ when $x = 1$ on the same pair of axes, for $-1 \leq x \leq 4$. Place a copy of the graphs in the space below.

c. Use the derivative to analyze the behavior of $f$.

As you respond to the following queries, keep in mind that you are dealing with two functions: $f(x) = x^2 + 3$ and $f'(x) = 2x$. The expression representing $f'$ can be used to calculate the derivative at any $x$-value where it exists. The value of the derivative provides information about the behavior of $f$. For instance, you can find the derivative when $x = 4$ by evaluating $f'$ at $x = 4$:

$$f'(4) = 2 \cdot 4 = 8$$

This tells you that the slope of the tangent line to the graph of $f(x) = x^2 + 3$ at $P(4, f(4))$—that is, at $P(4,19)$—is 8. Moreover, because the slope of the tangent line to the graph of $f$ is positive, you know that $f$ is increasing at $x = 4$.

(1) Fill in the entries in the following table. In each case, indicate what the value of $f'$ tells you about the behavior of the graph of $f$ (increasing, decreasing, levels off).

| $x$ | $f(x)$ | $f'(x)$ | Graph of f Increasing, Decreasing, or Level at P(x, f(x))? |
|---|---|---|---|
| $-3$ | | | |
| $-1$ | | | |
| $0$ | | | |
| $2$ | | | |
| $5$ | | | |

(2) Use the expression for $f'$ to find the set of all *x*-values where *f* is increasing—that is, where $f'(x) > 0$. Use interval notation to express your answer.

(3) Use the expression for $f'$ to find the set of all *x*-values where *f* is decreasing—that is, where $f'(x) < 0$.

**d.** Find the $(x, y)$-coordinates of the point on the graph where the slope of the tangent line is 5.

**e.** Sketch the graphs of the functions *f* and *f'* on the same pair of axes. Label the graphs.

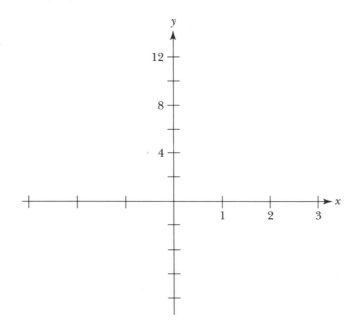

**3.** Consider $f(x) = 3x - 4$.

**a.** Use the definition of derivative to find an expression for $f'(x)$.

**b.** Use the expression for $f'(x)$ to find the rate of change of $f$ at $x = 12$.

**c.** Use the expression for $f'(x)$ to show that the graph of $f$ is always increasing.

**d.** The graph of $f$ has no turning points. Explain how your expression for $f'(x)$ supports this conclusion.

**e.** Graph the functions $f$ and $f'$ on the same pair of axes. Label the graphs

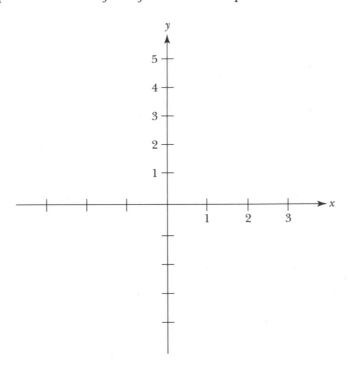

The derivative $f'$ of a function $f$ is a function. In the last task, you used the definition of derivative to represent $f'$ by an expression. The question in the next task is: What is the domain of this new function? You know that the domain of $f'$ is contained in the domain of $f$. But, are the two domains equal? Not necessarily. Just because a function is defined at $x = a$ does not guarantee that the tangent line at $P(a, f(a))$ exists. It is possible that $f(a)$ exists and $f'(a)$ does not, in which case the limit

$$\lim_{h \to 0} \frac{f(a + h) - f(a)}{h}$$

fails to exist. Thinking back to your study of limits, you know that:

• A limit does not exist if its left- and the right-hand limits exist, but have different values. Consequently, $f'(a)$ does not exist if the left and right limits, which are called the *left-* and *right-hand derivatives* exist, but have different values; that is, if

$$\lim_{h \to 0^-} \frac{f(a + h) - f(a)}{h} \neq \lim_{h \to 0^+} \frac{f(a + h) - f(a)}{h}$$

- A limit does not exist if its output values explode as the input values get closer and closer to the designated number. Consequently, $f'(a)$ does not exist if

$$\frac{f(a + h) - f(a)}{h}$$

blows up as $h$ approaches 0 from the left and/or the right; that is, if

$$\lim_{h \to 0^-} \frac{f(a + h) - f(a)}{h} = \pm\infty \quad \text{and/or} \quad \lim_{h \to 0^+} \frac{f(a + h) - f(a)}{h} = \pm\infty$$

In the next task, you will examine what goes wrong when you try to evaluate the limit at a place where the derivative does not exist. As you do the task, keep two things in mind: First, the following are equivalent ways of saying that $x = a$ is not in the domain of $f'$:

- $f$ is not differentiable at $x = a$.
- $f'(a)$ does not exist.
- $f'$ is undefined at $x = a$.

Second, as a line becomes more and more vertical, the slope of the line approaches $+\infty$ or $-\infty$. Consequently, the slope of a vertical line is undefined.

---

## Task 5-4: Inspecting the Domain of the Derivative

The domain of $f'$ is contained in the domain of $f$. What points are left out?

1. If the graph of a function $f$ has a sharp "peak" or "dip" at $x = a$, then $f'(a)$ does not exist. Examine why this is true.

   Actually, this is not a new idea. Remember the dude with the pointy hair in Unit 1? There were no tangent lines on the tips of his hair...

   a. Consider the graph of $f(x) = |x|$, which has a sharp dip at $x = 0$.

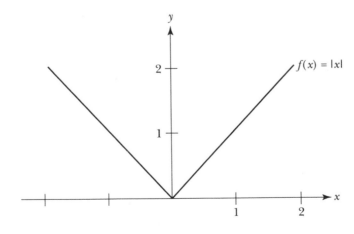

**(1)** Use a graphic approach to find the left-hand derivative of $f$ at $x = 0$.

    The *left-hand derivative* of $f$ at $x = 0$ is the limit of the slopes of the secant lines determined by $P(0, f(0))$ and $Q$, where $Q$ approaches $P$ from the left. Consequently, one way to determine the left-hand derivative of $f$ at $x = 0$ is to find the slope of the secant line determined by $P(0, f(0))$ and say $Q(-1, f(-1))$. Next, find the slope of the secant line for another value of $Q$ which is closer to $P$, but still to its left, for instance, $Q(-0.5, f(-0.5))$. Find the value of the left-hand derivative by generalizing your observations. Record your conclusion below.

The left-hand derivative of $f$ at $x = 0$ is _____ .

**(2)** Use a graphic approach to find the right-hand derivative of $f$ at $x = 0$.

    The *right-hand derivative* of $f$ at $x = 0$ is the limit of the slopes of the secant lines determined by $P(0, f(0))$ and $Q$, where $Q$ approaches $P$ from the right. Repeat the process you used to find the left-hand derivative in part (1), but this time consider a couple of values of $Q$ which are close to $P(0, f(0))$ but to the right of $P$. Record your conclusion below.

The right-hand derivative of $f$ at $x = 0$ is _____ .

**(3)** Explain why $f(x) = |x|$ is not differentiable at $x = 0$.

**b.** Generalize your observations.

    **(1)** Sketch an arbitrary function $f$ that is defined at $x = a$ and has a sharp peak or dip at $x = a$.

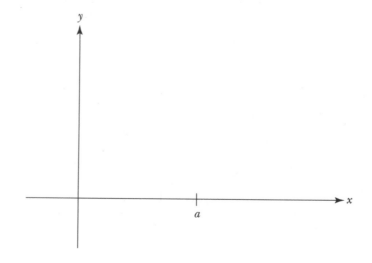

**(2)** Explain why $f$ is not differentiable at $x = a$.

2. If the graph of a function $f$ has a jump discontinuity at $x = a$, then $f'(a)$ does not exist. Examine why this is true.

   **a.** Consider the function

   $$f(x) = \begin{cases} x + 1, & \text{if } x \leq 2 \\ 5, & \text{if } x > 2 \end{cases}$$

   The graph of $f$ has a jump discontinuity at $x = 2$.

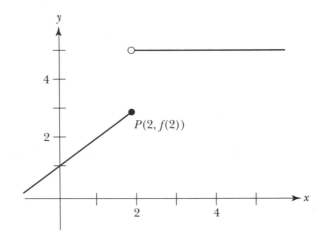

   **(1)** Find the value of the left-hand derivative of $f$ at $x = 2$ by calculating the slope of several secant lines determined by $P(2, f(2))$ and some points on the graph of $f$ to the left of $P$.

   The left-hand derivative of $f$ at $x = 2$ is _____.

   **(2)** Explain why the right-hand derivative of $f$ at $x = 2$ does not exist by sketching several secant lines determined by $P(2, f(2))$ and some points on the graph of $f$ to the right of $P$.

   **(3)** Explain why $f$ is not differentiable at $x = 2$.

**b.** Generalize your observations.

(1) Sketch an arbitrary function $f$ that has a jump discontinuity at $x = a$. Label the point $P(a, f(a))$ on your graph.

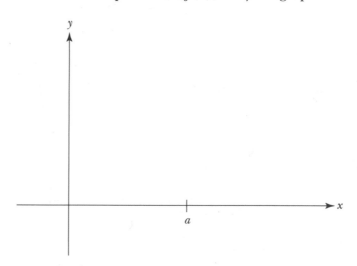

(2) Explain why $f$ is not differentiable at $x = a$. In the case of your function, which one-sided derivative does not exist?

**3.** If the graph of a function $f$ has a vertical tangent at $x = a$, then $f'(a)$ does not exist. Examine why this is true.

Consider the graph of the following function which has a vertical tangent at $x = a$.

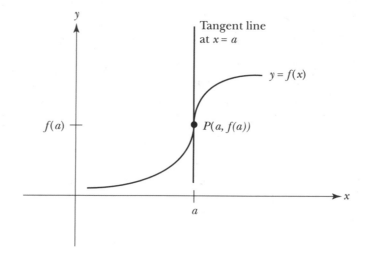

**a.** Explain why the left-hand derivative of *f* at *x* = *a* does not exist. Sketch the secant lines determined by $P(a, f(a))$ and several points on the graph of the function to the left of *P*. Describe what happens to the value of the slope of the secant line as the points get closer and closer to *P*.

**b.** Using a similar approach as in part a, explain why the right-hand derivative of *f* at *x* = *a* does not exist.

**c.** Explain why *f* is not differentiable at *x* = *a*.

In Unit 1, you used the motion detector to create distance versus time graphs, and you analyzed the behavior of the tangent line as it traveled along the graphs. You discovered that the value and sign of the slope of the tangent line provide a lot of information about the shape and behavior of the graph, including the following facts:

- The value of the slope of the tangent line corresponds to your velocity or instantaneous rate of change at a given time.
- The magnitude of the slope of the tangent line corresponds to the speed you are moving at a particular time.
- The sign of the slope indicates whether the function modeling your motion rises or falls from left to right. The sign is positive when the graph is increasing and negative when it is decreasing.
- If the value of the slope is zero, the graph of the function has a horizontal tangent.
- The change in the sign of the slope as the tangent line travels along the graph from left to right indicates the location of the function's local extrema. If the sign of the slope changes from positive to zero to negative, the function has a local maximum. If it changes from negative to zero to positive the function has a local minimum.
- If the graph lies above the tangent line, the function is concave up. If it lies below, it is concave down.

In the next task, you will reexamine these ideas and express them in terms of the derivative. You will also investigate how the graph of the derivative is related to the graph of the given function. In particular, based on behavior of *f*, you will determine the following:

- The places where *f'* is undefined
- The places where the graph of *f'* intersects the horizontal axis

> • The intervals where the graph of $f'$ lies above the horizontal axis and the intervals where the graph lies below

As you explore the relationship between $f$ and $f'$, you will develop the statement of the First Derivative Test.

## Task 5-5: Investigating the Relationship Between a Function and Its Derivative

1. Summarize what you know about the relationship between a function and the tangent line to its graph. Express these ideas in terms of the derivative and describe their impact on the graph of the derivative.

   **a.** Consider the cases where a function $f$ is increasing, decreasing, or has a horizontal tangent at $x = a$.

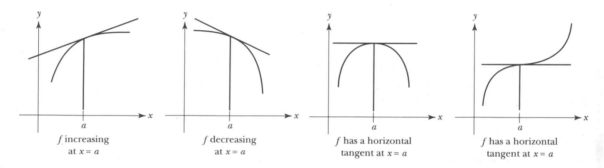

| | | | |
|---|---|---|---|
| $f$ increasing at $x = a$ | $f$ decreasing at $x = a$ | $f$ has a horizontal tangent at $x = a$ | $f$ has a horizontal tangent at $x = a$ |

   Summarize what's happening in each of these situations by filling in the following chart. The first row has been completed for you. Look it over carefully before completing the other rows.

| Description of $f$ at $x = a$ | Shape of Graph of $f$ near $x = a$ | Sign of Slope of TL at $x = a$. i.e., Sign of $f'(a)$ | Location of Graph of $f'$ at $x = a$ |
|---|---|---|---|
| $f$ increasing | Rises from left to right | + | Above horizontal axis |
| $f$ decreasing | | | |
| $f$ has a horizontal tangent | | | |

*Note: TL stands for "tangent line."*

**b.** Consider the cases where a function *f* has a local minimum, a local maximum, or a place where the graph levels off at *x = a*.

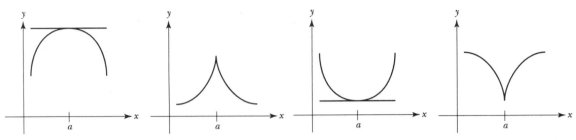

*f* has a local maximum at *x = a*            *f* has a local minimum at *x = a*

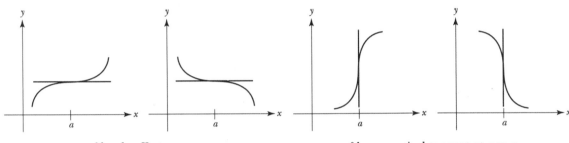

*f* levels off at *x = a*            *f* has a vertical tangent at *x = a*

Summarize what's happening in each of these situations by filling in the following chart. The first row has been completed for you. As usual, look at it carefully before completing the other rows.

*Note: DNE stands for "does not exist."*

| Description of *f* at *x = a* | Shape of Graph of *f* near *x = a* | Change in Sign of Slope of TL near *x = a*. i.e., Change in Sign of *f'(x)* | Behavior of Graph of *f'* near *x = a* |
|---|---|---|---|
| *f* has a local maximum | Increases to left of *x = a* and decreases to right. | Changes from + to 0 at *x = a* to −, or from + to DNE at *x = a* to −. | Falls from left to right. Crosses axis at *x = a*. Either 0 or undefined at *x = a*. |
| *f* has a local minimum | | | |
| *f* levels off | | | |
| *f* has a vertical tangent | | | |

2. Suppose $f$ is a an arbitrary function. How are $f$ and $f'$ related? What does $f'$ tell you about the graph of $f$? What does $f$ tell you about the behavior of $f'$?

Consider the graph of $f$.

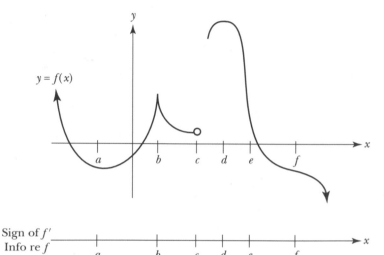

a. Fill in the Info re $f$ (information regarding $f$) on the chart given below the graph.

   (1) Label the intervals where $f$ is increasing (inc) and decreasing (dec).

   (2) Mark the $x$-values where $f$ has a local maximum (LM), has a local minimum (lm), and levels off (LO).

b. Investigate the relationship between the shape of the graph of $f$ and the sign of $f'$ as you fill in the sign chart for $f'$.

   (1) Find all the $x$-values where $f'(x)$ does not exist. Write DNE above each value on the sign chart for $f'$. Note that these are the $x$-values where $f'$ is undefined.

   (2) Find all the $x$-values such that $f'(x) = 0$, that is, where the slope of the tangent line to the graph of $f$ is 0. Mark the values with a 0 on the sign chart for $f'$. Note that these are the $x$-values where the graph of $f'$ intersects the horizontal axis.

   (3) Identify all the intervals where $f'(x) < 0$, that is, where the slope of the tangent line to the graph of $f$ is negative since $f$ is decreasing. Place a string of $-$'s above each interval on the sign chart for $f'$. Note that these are the intervals where the graph of $f'$ lies below the horizontal axis.

   (4) Identify all the intervals where $f'(x) > 0$, that is, where the slope of the tangent line to the graph of $f$ is positive since $f$ is increasing. Place a string of $+$'s above each interval on the sign chart for $f'$. Note that these are the intervals where the graph of $f'$ lies above the horizontal axis.

**(5)** Summarize what you know about the graph of $f'$.

What if you do not know what the graph of $f$ looks like, but you do know the sign chart for $f'$? How can you locate the local extrema by examining the sign chart for $f'$? The First Derivative Test, which follows directly from the observations you made in the last task, provides you with a way of identifying local extrema.

## The First Derivative Test

Suppose $f'(c) = 0$ or $f'(c)$ does not exist. Suppose $I$ is an open interval containing $c$, where $f$ is a continuous and differentiable everywhere on $I$, except possibly at $x = c$.

    **i.** If $f'(x) > 0$ for all $x$ immediately to the left of $c$ and $f'(x) < 0$ for all $x$ immediately to the right of $c$—that is, if $f$ is increasing to the left of $c$ and decreasing to the right of $c$—then $f$ has a local maximum at $x = c$.

    **ii.** If $f'(x) < 0$ for all $x$ immediately to the left of $c$ and $f'(x) > 0$ for all $x$ immediately to the right of $c$—that is, if $f$ is decreasing to the left of $c$ and increasing to the right of $c$—then $f$ has a local minimum at $x = c$.

The sign of the derivative of a function gives you a lot of information about the shape of the graph of its underlying function. If the derivative is positive over some interval, you know that the function is increasing over that interval. If the derivative is negative over some interval, the function is decreasing over that interval. If the derivative equals 0, the function has a horizontal tangent—which is either a local extremum or a leveling off point.

In addition to indicating the sign of the derivative on the sign chart, you can also indicate values where the derivative does not exist. This, too, provides you with information about the behavior of the graph of the underlying function. In particular, if the function is continuous at the point where the derivative does not exist, the graph of the function may have a cusp or a vertical tangent. On the other hand, if the function is not continuous at the point where the derivative does not exist, the graph of the function may have a jump.

In the following activities, you will interpret some sign charts for derivatives and see what they tell you about the shapes of the graphs of the underlying functions.

## Task 5-6: Gleaning Information About the Graph of a Function from Its Derivative

**1.** Consider the following sign chart:

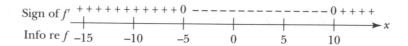

Sign of $f'$  $+ + + + + + + + + + + 0\ {-}\ {-}\ {-}\ {-}\ {-}\ {-}\ {-}\ {-}\ {-}\ {-}\ {-}\ {-}\ 0 + + + +$

Info re $f$  $-15 \quad -10 \quad -5 \quad 0 \quad 5 \quad 10$

   Assume $f$ is defined for all real numbers. In addition, assume that the only zeros of $f'$ are at $x = -5$ and $x = 10$. Use the sign chart for $f'$ to answer the following questions about the shape of the graph of $f$.

**a.** Determine the intervals where $f$ is increasing (inc). Record this information regarding $f$ on the sign chart given above.

**b.** Determine the intervals where $f$ is decreasing (dec). Record this information regarding $f$ on the sign chart given above.

**c.** Find all the $x$-values where $f$ has a local maximum (LM) or a local minimum (lm) and where the graph of $f$ levels off (LO). Record this information regarding $f$ on the sign chart given above.

**d.** Sketch a graph of a function $f$ whose derivative has the given sign chart.

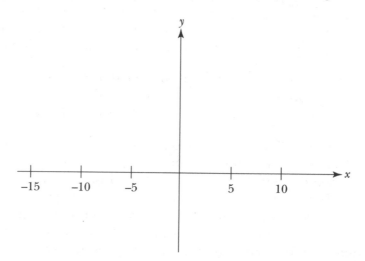

**2.** Consider the following graph of the derivative $f'$:

Graph of $f'$

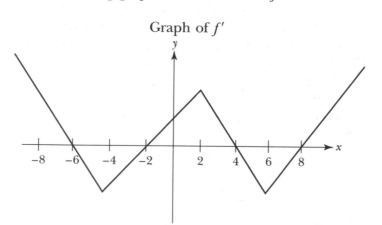

**a.** Find the associated sign chart for $f'$.

**b.** Describe the shape of the graph of $f$ by filling in information regarding $f$; that is, indicate the intervals where $f$ is increasing and decreasing, and give the location of all the extrema of $f$.

**c.** Suppose the graph of $f$ lies above the horizontal axis for $-8 \le x \le 8$; that is, $f(x) > 0$ for $-8 \le x \le 8$. Sketch a graph of a function $f$, where the graph of $f'$ is given above.

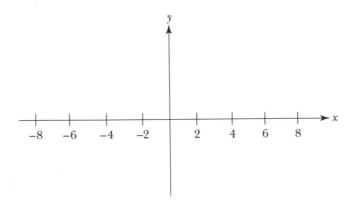

**3.** On the axes given below, sketch the general shape of the graph of a function $f$ which satisfies the following conditions:

- $f$ defined for all real numbers
- $f'(x) = x^2$
- $f(0) = 3$

*Note: First, find the sign chart for the derivative f' and fill in the "info re f."*

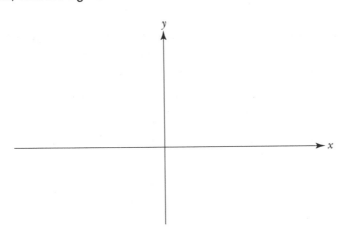

## Unit 5 Homework After Section 1

- Complete the tasks in Section 1. Be prepared to discuss them in class.

- Calculate some derivatives using the definition in HW5.1. Use the derivative to analyze the function. Compare the graphs of a function and its derivative.

**HW5.1** Recall that, by definition,

$$f'(x) = \lim_{h \to 0} \frac{f(x+h) - f(x)}{h}$$

provided the limit exists.

1. For each of the following functions:

   **i.** Use the definition to find its general derivative.

   **ii.** Sketch the graphs of $f$ and $f'$ on the same pair of axes. Label the graphs.

   **a.** $f(x) = 6$

   **b.** $f(x) = 6.5x - 10$

   **c.** $f(x) = x^3$. Note: $f(x+h) = (x+h)^3 = x^3 + 3x^2h + 3xh^2 + h^3$

2. Suppose $f(x) = x^2 - 2x + 5$.

   **a.** Use the definition to find $f'(x)$.

   **b.** Sketch the graphs of $f$ and $f'$ on the same pair of axes. Label the graphs.

   **c.** Use $f'(x)$ to find the equation of the tangent line at $P(2, f(2))$.

   **d.** Use the expression for $f'(x)$ to find the $(x, y)$-coordinates of the point on the graph of $f$ where the tangent line is horizontal to the graph.

   **e.** Use the expression for $f'(x)$ to find the intervals where the function is increasing and the intervals where it is decreasing.

3. Use the definition of derivative to show that each of the following statements is true. Support the conclusion of the statement by graphing a typical $f$ and its derivative $f'$ on the same pair of axes. Label the graphs.

   **a.** The derivative of a constant function is always 0; that is, if $f(x) = c$, where $c$ is a fixed constant, then $f'(x) = 0$ for all $x$.

   **b.** The derivative of a linear function is the slope its graph; that is, if $f(x) = mx + b$, where $m \neq 0$, then $f'(x) = m$ for all $x$.

- Use the derivative to analyze the shape of an underlying function in HW5.2.

**HW5.2** Sketch the graph of a function that satisfies the given condition.

**1.** Sketch the graph of a function $f$ satisfying the following conditions:

- $f$ is defined for all real numbers.
- The sign chart of the derivative of $f$ is

where DNE indicates that the derivative does not exist.

**2.** Sketch the graph of a function $s$ satisfying the following conditions:

- $s$ is continuous for all real numbers.
- $s$ is differentiable everywhere, except at $t = 6$—that is $s'(6)$ DNE.
- $s'(t) \leq 0$ for all $t$, except $t = 6$.
- $s'(3) = s'(9) = 0$.
- $s(6) = 0$.

**3.** Sketch the graph of a function $h$ satisfying the following conditions:

- $h$ is continuous for all real numbers.
- $h'(x) = x^2 - x$.
- The graph of $h$ passes through the point $P(\frac{1}{2}, -4)$.

**4.** Sketch the graph of a function $f$ satisfying the following conditions:

- $f$ is continuous for all real numbers.
- $f(x) \leq 0$, for all $x$.
- $f(10) = 0$.
- The graph of $f'$ has the following shape:

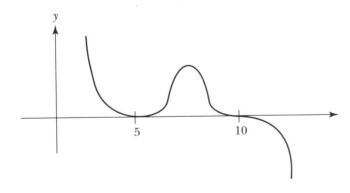

- Compare the graphs of a function and its derivative in HW5.3.

   **HW5.3** Each of the following graphs displays the graph of a function $f$ and its derivative $f'$. Indicate which graph is which. Justify your choices.

**1.**

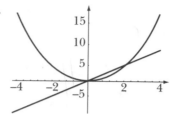

**2.**

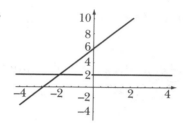

**3.**

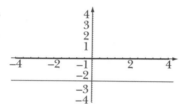

**4.**

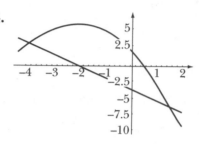

**5.**

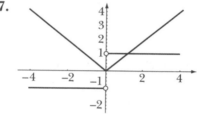

**6.**

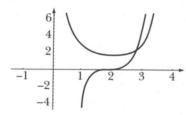

**7.**

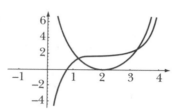

**8.**

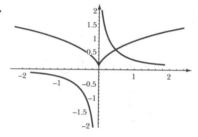

- Many different functions have the same derivative. Make a first pass at determining the relationship between two functions that have the same derivative in HW5.4.

**HW5.4** Keep in mind that the value of $f'$ can be determined by the value of the slope of the tangent line as it travels along the graph of $f$.

1. Suppose the graph of $f'$ is given below:

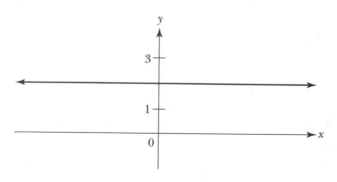

**a.** Based on the shape of the graph of $f'$, explain why the graph of $f$ is a line.

**b.** Give the slope of the graph of $f$.

**c.** Assume $f$ passes through the point $P(0,0)$.

   **(1)** Represent $f$ by a graph. Label your axes.
   **(2)** Represent $f$ by an expression.

**d.** Instead of assuming the graph of $f$ passes through the point $P(0,0)$, assume the graph of $f$ passes through the point $P(0,-3)$.

   **(1)** Represent $f$ by a graph. Label your axes.
   **(2)** Represent $f$ by an expression.

**2.** Suppose the graph of $f'$ is given below:

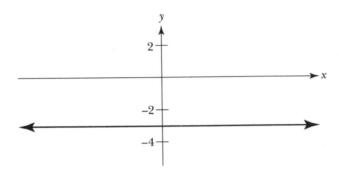

**a.** Assume $f(0) = 2$. Represent $f$ by a graph and an expression.

**b.** Assume $f(2) = 0$. Represent $f$ by a graph and an expression.

**3.** What's going on here? In each instance, there are at least two functions that have the same derivative. Describe the relationship between the graphs of linear functions that have the same derivative.

• The value of $f'(a)$ equals the slope of the tangent line to the graph of $f$ at $x = a$. It also gives the rate of change of $f$ at $x = a$. With this interpretation, instead of finding the limit of the slopes of secant lines, you think about taking the limit of average rates of change. Calculate some average rates of change in HW5.5.

**HW5.5** Suppose $b$ is a point near $a$ in the domain of $f$. Then

*average rate of change* of $f$ between $x = a$ and $x = b$

   $= $ slope of secant line determined by $P(a,f(a))$ and $Q(b,f(b))$

   $= \dfrac{f(a) - f(b)}{a - b}$

   $= \dfrac{\text{change in output values}}{\text{change in input values}}$

whereas

(*instantaneous*) *rate of change* of $f$ at $x = a$

= slope of the tangent line to the graph of $f$ at $P(a, f(a))$

$$= \lim_{b \to a} \frac{f(a) - f(b)}{a - b}$$

$$= f'(a)$$

Calculate some average rates of change.

1. Fruit flies multiply fast. The following graph of the function $N$ shows the growth of a fruit fly population during a 5-hour period.

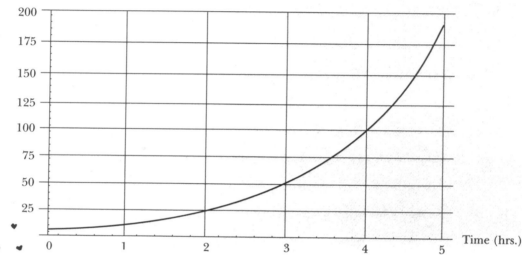

Fruit flies

a. Approximate the numerical value for the average rate in the fruit fly population over the following time intervals:

(1) Between 3 and 5 hours
(2) Between 3 and 4 hours
(3) Between 1 hour, 30 minutes and 3 hours

b. Give a graphic representation of each of the numerical values which you found in part a by drawing the associated secant line on the graph of $N$. Label the lines (1), (2), and (3) to indicate which time interval the line represents.

2. The following table shows the growth chart for Mark:

| Age (yr.) | 0 | 2 | 4 | 6 | 8 | 10 | 12 | 14 | 16 |
|-----------|----|----|----|----|----|----|----|----|----|
| Height (in.) | 19 | 23 | 28 | 36 | 46 | 55 | 66 | 70 | 72 |

**a.** Graph Mark's growth (either by hand or using your CAS).

**b.** Describe Mark's growth rate during the 16-year period.

**c.** Find Mark's average rate of growth in inches per year from the time he was born until he was 4 years old.

**d.** Find Mark's average rate of growth in inches per year from age 12 to age 16.

**e.** Find the 2-year period during which time Mark had his largest average rate of growth.

**f.** Find the four year period during which time he had his slowest average rate of growth.

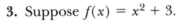

**3.** Suppose $f(x) = x^2 + 3$.

**a.** Find the average rate of change of $f$:

    **(1)** Between $x = 1$ and $x = -1$
    **(2)** Between $x = 1$ and $x = 0$
    **(3)** Between $x = 1$ and $x = 2$
    **(4)** Between $x = 1$ and $x = 3$

**b.** Graph the function $f$ (either by hand or using your CAS) for $-1 \le x \le 3$.

**c.** On your graph, carefully sketch the secant lines determined by the pairs of input values given in part a. Label each secant line on your graph and indicate its slope.

• In the case of the distance function, the derivative gives the velocity of an object at any given time. Apply this interpretation of derivative as you do HW5.6.

**HW5.6** Time is running out. Suppose you have 15 seconds to get to your destination 30 meters away. You start off at an acceptable pace, but because you are loaded down with books, you keep slowing down. The graph below shows your velocity function for the next 15 seconds.

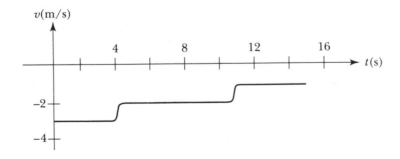

1. Give a verbal description of your movement. For instance, you might say, "During the first 4 seconds my velocity is −3 meters/second. Therefore, during the first 4 seconds, I am decreasing my distance from my destination, moving at a rate of 3 meters each second," and so on.

2. Approximate the total distance you move during the 15-second time period.

3. Recalling that initially you are 30 meters from your destination, sketch your underlying distance function.

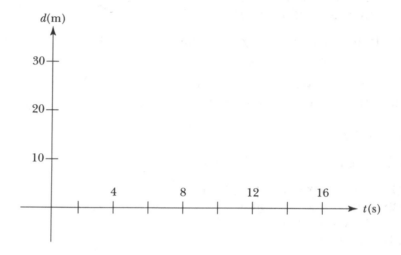

- Think about the meaning of the Mean Value Theorem for Derivatives in HW5.7.

**HW5.7** According to the *Mean Value Theorem for Derivatives*:

If $y = f(x)$ is continuous on the closed interval $[a,b]$ and is differentiable at every point in the open interval $(a,b)$, then there is at least one point $c$ between $a$ and $b$ at which

$$\frac{f(b) - f(a)}{b - a} = f'(c)$$

1. Explain why the Mean Value Theorem for Derivatives makes sense. Develop your explanation as follows:

   a. Sketch an appropriate diagram.

      (1) Sketch a pair of axes, and label $a$ and $b$ on the horizontal axis where $a < b$.
      (2) Sketch the graph of a "smooth" function $f$ over $[a, b]$—that is, a function which is continuous on the closed interval $[a, b]$ and is differentiable at every point in the open interval $(a, b)$.
      (3) Label $f(a)$ and $f(b)$ on the vertical axis.
      (4) Label the points $A(a, f(a))$ and $B(b, f(b))$. Draw a line through $A$ and $B$.

**b.** Answer some questions.

**(1)** What information does the value of

$$\frac{f(b) - f(a)}{b - a}$$

give you about the line determined by the points *A* and *B*?

**(2)** If *c* is *any* point between *a* and *b*, what information does the value $f'(c)$ give you about the tangent line to the graph of *f*?

**c.** Show that the conclusion to the Mean Value Theorem for Derivatives holds.

**(1)** Find *c* between *a* and *b* so that

$$\frac{f(b) - f(a)}{b - a} = f'(c)$$

**(2)** Label $f(c)$ on the vertical axis. Sketch the tangent line at $C(c, f(c))$.
**(3)** Describe the relationship between the line connecting the points *A* and *B* and the tangent line at *x* = *c*.

**2.** Apply the Mean Value Theorem for Derivatives to some real-life situations.

**a.** You stop at a red light. When the light turns red, you start off slowly, continually increasing your speed. After 15 seconds you have traveled a quarter of a mile, where your motion is depicted in the following distance versus time graph:

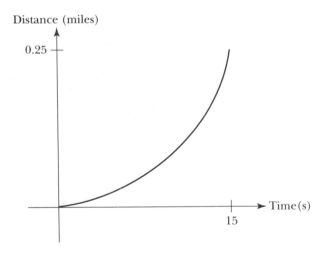

**(1)** Show that between 0 and 15 seconds your average speed is 60 MPH. *Hint*: 15 seconds = $\frac{1}{240}$ hours.
**(2)** On the graph, sketch the secant line whose slope corresponds to your average speed during the 15-second time interval.

(3) Note that your speed at any given moment corresponds to the slope of the tangent line at that point. According to the Mean Value Theorem for Derivatives, at some time $c$ between 0 and 15 seconds your speed was 60 MPH. Label $c$ on the $t$-axis. Sketch the tangent line whose slope is 60.

(4) The posted speed limit is 60 MPH. A state trooper pulls you over at 15 seconds. Do you deserve a ticket? Justify your response.

**b.** You work at a pretzel factory, where you make 1,200 pretzels during your 8-hour shift. The number of pretzels you can produce in $t$ hours is given by the production function $Q(t)$, whose graph is given below:

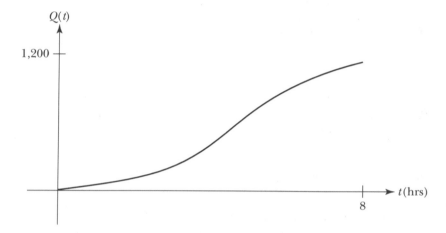

(1) Find your average rate of productivity between 0 and 8 hours.

(2) On the graph, sketch the secant line whose slope corresponds to your average rate of productivity during the 8-hour shift.

(3) According to the Mean Value Theorem for Derivatives, there exists at least one time $c$ during your shift when your productivity rate equals your average rate of production. In your case, there are two choices for $c$, say, $c_1$ and $c_2$. Label $c_1$ and $c_2$ on the $t$-axis. Sketch the associated tangent lines.

(4) Label the time on the $t$-axis when your rate of productivity is the greatest. Sketch the associated tangent line.

(5) The moment of maximum efficiency, which you labeled in part (4), is called your *point of diminishing returns*. Explain why this terminology makes sense.

• Describe your understanding of derivative in HW5.8.

**HW5.8** What is a derivative? Write a short essay explaining what a derivative is. In your essay:

i. List the various ways to interpret the value of a derivative.

ii. Give a formal definition of "derivative."

iii. Explain why $f'(a)$ gives the slope of the tangent line to the graph of $f$ at $P(a, f(a))$. Support your explanation with a diagram. Label your diagram.

iv. Describe conditions under which the derivative does not exist.

**v.** Summarize how you can use the derivative to gain information about the behavior of its underlying function.

- Explore the use of sigma notation in HW5.9. You will use this notation in the next section.

**HW5.9** Many times throughout your study of mathematics, you will be asked to find the sum of a sequence of numbers. For example, suppose your credit card company has a strict policy against late payments. For each day your payment is late, that many dollars will be added onto your total bill in addition to the amount added on for the previous days. In other words, if you are three days late, $6 will be added onto your total bill—that is, $3 for the third day plus $2 for the second day and $1 for the first day.

When there are only three terms to be added together, it is easy to write out $1 + 2 + 3$, but what if you are 100 days late? It's tedious to write out $1 + 2 + 3 + 4 + 5 + \cdots + 100$. Consequently, a special notation called *sigma notation* has been developed to represent sums of numbers. Using sigma notation, you can represent

$$1 + 2 + 3 + 4 + 5 + \cdots + 100 \text{ as } \sum_{i=1}^{100} i$$

The capital Greek letter sigma $\Sigma$ represents a sum and the variable $i$ is the *index* of summation. In this case, the values of $i$ range through the integers from 1 to 100. The limits on the index will vary depending on your problem, as will the expression to be summed. For instance,

$$\sum_{i=12}^{16} i^2 \text{ represents } 12^2 + 13^2 + 14^2 + 15^2 + 16^2.$$

1. Rewrite each summation as the sum of a sequence of numbers. Do not calculate the actual value of the sum.

   **a.** $\displaystyle\sum_{i=3}^{7} (2i + 1)$      **c.** $\displaystyle\sum_{i=-3}^{2} (-i + 3)$

   **b.** $\displaystyle\sum_{i=1}^{4} i^3$      **d.** $\displaystyle\sum_{i=0}^{5} \frac{i}{i + 1}$

2. Represent the following sums using sigma notation.

   **a.** $2 + 3 + 4 + \cdots + 12$      **c.** $1 + \dfrac{1}{2} + \dfrac{1}{3} + \dfrac{1}{4} + \dfrac{1}{5} + \cdots + \dfrac{1}{12}$

   **b.** $1 + 4 + 9 + 16 + \cdots + 81$      **d.** $15 + 20 + 25 + \cdots + 65$

3. Because calculating these long sums can be very tedious, it is helpful to let the computer do it for you. In ISETL, the %+ symbol is used to represent $\Sigma$. For example, to find the value of

$$48 + 75 + 108 + \cdots + 300$$

or

$$\sum_{i=4}^{10} 3i^2$$

you can enter in ISETL

$$\%+ \ [3*i**2 : i \text{ in } [4..10]];$$

where

> $\%+$ tells ISETL to sum a sequence of values
> $3*i**2$ represents the values to be summed
> $i$ is the index of summation
> 4 and 10 are the lower and upper bounds on $i$

Look again at the sums you considered in parts 1 and 2. In each case,

**i.** Translate the mathematical notation for the sum into ISETL.
**ii.** Use ISETL to find the value of the sum.

Print a copy of your ISETL session. Label your results.

SECTION 2

## The Definite Integral

The definite integral is one of the most important and fundamental concepts in calculus. It is extremely useful for measuring quantities and determining how quantities accumulate. For example, definite integrals can be used to help a maintenance worker (who has taken calculus!) figure out how much rubber sealer he will need to coat the bottom of a kidney shaped swimming pool; or the president of Chemical Bank can use a definite integral to determine his accumulated net assets on a particular day, based on the rate at which money has been deposited or withdrawn from the bank up to that time. Surprising as it may seem, these two types of calculations—finding the area of an oddly shaped region and determining an accumulation resulting from a variable rate of change—are very closely connected mathematically.

In order to be able to identify situations where it is appropriate to use a definite integral, you need to develop a conceptual understanding of what a definite integral represents. This is the primary objective of this section. The tasks have been designed to help you discover its definition and develop a mental image associated with the concept of definite integral.

In this section, you will examine the relationship between definite integrals and areas. You will use ISETL to do some of the calculations and to help you understand the mathematical notation underlying the definition of definite integral. You will explore ways to approximate areas and accumulations and think about how to improve your approximations, making them more exact. You will develop a general method which applies in all cases. Finally, you will make a first pass at examining the connection between definite integrals and derivatives.

Although this initial encounter with the concept of a definite integral will involve using a definite integral to represent the area or a region, it can be used to represent numerous quantities that accumulate. In the following units, you will return to your study of integrals and explore other situations where the idea is applicable. In addition, you will develop a straightforward way to find exact solutions to these types of problems (without evaluating a limit).

Your study of definite integrals begins as you approximate some areas in the next three tasks.

---

## Task 5-7: Finding Some Areas

Should your client buy this property? You have graduated from college after distinguishing yourself in this course and are working for a team of real estate consultants. One of your clients is considering buying some vacation property that borders on the Susquehanna River. The property comes equipped with a boat house and dock, but there are a few problems with them. The dock needs to be rebuilt, the back of the boat house that faces the river is in desperate need of paint, and the survey information has been lost. Your client needs to know the areas of the  dock and the back of the boathouse (so she can calculate the cost of refurbishing them), and she needs to determine the area of the property (so she can determine if she can afford to buy it).

1. The dock has good supports, but the wood on the deck part is rotten. The dock deck is 4 feet wide and 6 feet long. Do a scale drawing of the deck and find its total area.

2. Suppose the local lumber mill sells leftover treated outdoor boards which are exactly 4" wide, 2" deep, and 4' long. How many boards will be needed to rebuild the deck? Sketch the boards needed on your drawing above.

3. If the treated outdoor boards cost $9.50 each and your labor is free, how much will it cost to refurbish the deck on the dock?

4. The back of the boat house is depicted in the diagram below. How many square feet need to be painted? Explain how you figured this out.

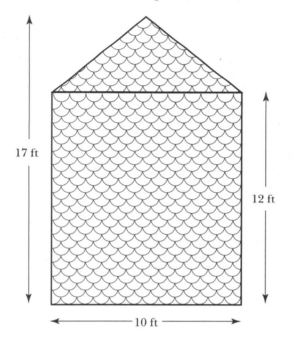

17 ft

12 ft

10 ft

5. If a quart of special preservative costs $12.25 and covers 50 square feet. How much will the preservative cost to refurbish the back face of the boat house?

6. The seller has made a careful sketch of the site boundaries on graph paper, where each square in the grid is 50' × 50', as shown below. The property is bounded on the north by the Susquehanna River and on the south by Rt. 70. Approximately how many square feet are included in the property? Describe briefly the techniques you thought about using to find the answer, and then explain why you chose the one you used.

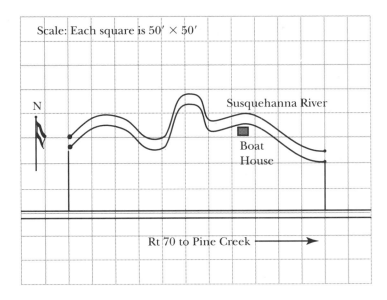

7. If the property is being sold for 1 penny per square foot, approximately how much is the seller asking for this plot of land?

8. After paying your consulting fee, your client can only afford to pay a total of $1200 for the property and the lumber and paint to fix things up. Can she afford to buy it?

In part 6 of the last task, you approximated the area of the region bounded by the Susquehanna River and Route 70. Chances are that you did this by counting the number of squares in the grid covering the region and then summing the areas of the squares. Explore some other possible ways to do this in Task 5-8.

## Task 5-8: Describing Some Possible Approaches

1. First, you know how to find the areas of a variety of regular shapes. List the shapes whose areas you can calculate, and give the associated area formulas.

2. Suppose you want to find the area of an oddly shaped region lying under a given curve and over a closed interval, such as in Figure 1.

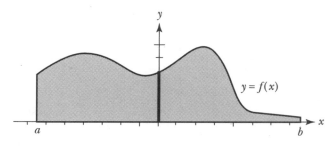

Figure 1.

For each regular shape, which you listed in part 1, make a rough sketch of the region in Figure 1. Cover the region with pieces having the desired shape. For instance, cover the region with circles. The pieces can touch, but cannot overlap. Observe that you can approximate the area of the region by summing the areas of the pieces.

There are a number of ways to approximate the area of a region under a curve and over a given interval. In the next few tasks, you will examine a rectangular approach, which involves covering the region with rectangles, formed by using the right end-point of each subinterval. To use this approach:

- Partition the given interval into equal size pieces or subintervals.
- For each subinterval, form a rectangle whose base is the subinterval and whose height is the value of the function at the right end-point of the subinterval.
- Calculate the area of each rectangle.
- Approximate the area under the curve by summing the areas of the rectangles.

Apply this approach to approximate the area of the region bounded by the Susquehanna River which you considered in Task 5-7, part 6.

## Task 5-9: Applying a Rectangular Approach

*Note: A blowup of the region under consideration appears on the next page.*

1. Place the region on a coordinate system, with the southwest corner of the property at the origin and the horizontal and vertical axes determined by the highway and the western boundary respectively. Carefully label the axes in terms of feet.

2. Suppose $h$ is the function whose value at points along the southern boundary is determined by measuring the distance from the highway to the riverbank. For example, if you look carefully at the diagram, $h(0)$ is approximately 131 feet. Approximate the value of $h$ at each of the following points along the southern boundary:

$$h(0) = 131'$$
$$h(25) =$$
$$h(100) =$$
$$h(275) =$$
$$h(350) =$$

3. Suppose you wanted to approximate of the area of the property by dividing the property into 11 rectangular strips of equal widths, running north to south. A mathematician refers to the base of each strip as a *subinterval*. Find the size of each subinterval.

4. Assume the height of each rectangular strip is determined by the value of $h$ at the *right* end-point of the associated subinterval.

   a. Sketch the 11 rectangles on the blowup of the region.

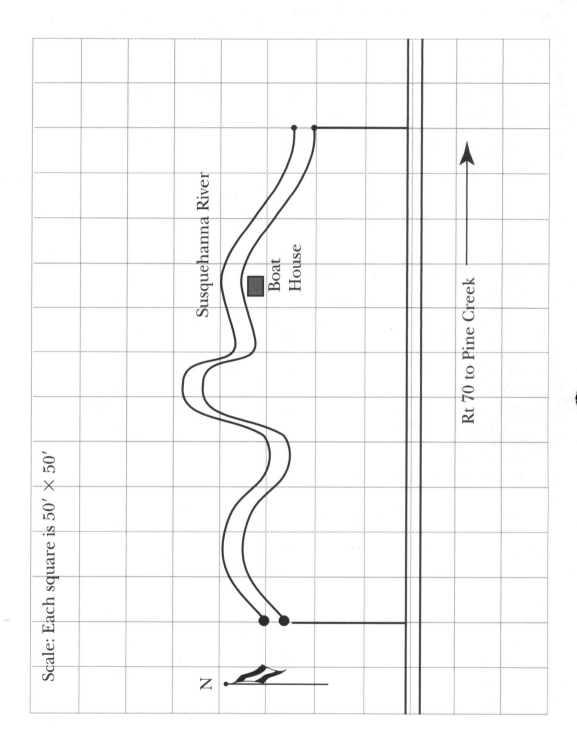

Scale: Each square is 50′ × 50′

**b.** First make sure that the column headings on the chart given below make sense to you, then fill in the chart.

| Subinterval # $i$ | Right End-Point of $i$th Subinterval $x$(ft) | Height of $i$th Rectangle $h(x)$ (ft) | Width of $i$th Rectangle $\Delta x$ (ft) | Area of $i$th Rectangle $h(x) \cdot \Delta x$ (ft$^2$) |
|---|---|---|---|---|
| 1 | 50 | | | |
| 2 | 100 | | | |
| 3 | 150 | | | |
| 4 | 200 | | | |
| 5 | 250 | | | |
| 6 | 300 | | | |
| 7 | 350 | | | |
| 8 | 400 | | | |
| 9 | 450 | | | |
| 10 | 500 | | | |
| 11 | 550 | | | |

**c.** Use the information on the chart to approximate the area of the region. How does this result compare to the approximation you found in Task 5-7, part 6?

**d.** Describe how you can make your approximation more precise.

In the last task, you approximated the area of a region by using a small number of rectangles. Your goal is to find the actual value. Intuitively, it should make sense that the more rectangles you consider, the more accurate your approximation will be. In the next task, you will develop the mathematical notation for the general situation. You will represent the approximate area of a region by an expression and then pass to the limit to represent the exact area.

## Task 5-10: Considering the General Situation

Suppose $f$ is a non-negative, continuous function, where $a \le x \le b$. Consider the region under the graph of $f$ and over the closed interval $[a,b]$. Approximate the area of this region by covering it with $n$ rectangles and then summing the areas of the rectangles. Express the exact area in terms of a limit.

1. Sketch the graph of $f$, for $a \le x \le b$. Make the graph of $f$ curvaceous and make it lie above the $x$-axis.

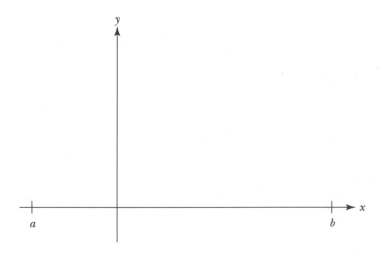

2. Partition the closed interval $[a,b]$ into $n$ equal subintervals and find a formula for the right end-point of each subinterval. These subintervals will form the bases of the rectangles that you will use to approximate the area of the region.

   **a.** Observe that the length of the entire interval is $b - a$. Find a formula for the width of each of the $n$ subintervals in terms of $a$, $b$, and $n$. Call this width $\Delta x$—which is read "delta $x$" and denotes a small change in $x$.

   $$\Delta x =$$

   **b.** Find formulas for the end-points of the $n$ subintervals and give each end-point a name.

   The initial end-point is $a$. Call this $x_0$; that is,

   $$x_0 = a$$

   **(1)** The right end-point of the first subinterval is $\Delta x$ away from $a$. Find a formula for the end-point of this subinterval in terms of $a$ and $\Delta x$. Call this end-point $x_1$.

   $$x_1 =$$

   **(2)** The end-points of the successive subintervals are $\Delta x$ apart. Find formulas for the right end-points of the second and third subintervals in terms of $a$ and a multiple of $\Delta x$. Call the end-points of these subintervals $x_2$ and $x_3$, respectively.

$$x_2 =$$

$$x_3 =$$

(3) Recall that $n$ represents the total number of subintervals. You know that $n$ is a positive integer, but you don't know which one. Consequently, it is impossible to write out all the values of the right end-points of the subintervals. What might you do in this case? One approach is to consider an arbitrary subinterval—that is, instead of referring to the first, second, or third subinterval, refer to the $i$th subinterval where $1 \le i \le n$, as there are $n$ subintervals.

Use the pattern that emerged when you found formulas for $x_1$, $x_2$, and $x_3$ to find a formula for the right end-point of the $i$th subinterval. As before, express the formula in terms of $a$ and a multiple of $\Delta x$. Call the end-point of the $i$th subinterval $x_i$.

$$x_i = \qquad\qquad , \quad \text{where } 1 \le i \le n$$

(4) The right end-point of the last, or $n$th, subinterval is $b$. Find a formula for the right end-point of this subinterval in terms of $a$ and a multiple of $\Delta x$.

$$x_n = b =$$

c. Mark the values of $x_i$, where $0 \le i \le n$, on the $x$-axis on your diagram in part 1.

(1) Relabel $a$ as $x_0$ and $b$ as $x_n$.

(2) Starting at the value for $a$, make three, equally spaced tick marks on the $x$-axis, representing the right end-points of the first three subintervals. Label them $x_1$, $x_2$, and $x_3$.

(3) Because it's impossible to label the end-point of every subinterval, after the tick mark for $x_3$, put three dots on the $x$-axis to indicate that some points are missing.

(4) Make tick marks for $x_{i-1}$ and $x_i$, which represent the left and right end-points of the $i$th subinterval, and label them appropriately.

(5) After the tick mark for $x_i$, put three more dots on the $x$-axis to indicate that more points are missing.

(6) Finally, noting that $b$ is $x_n$, make a tick mark for $x_{n-1}$, which is the left end-point of the last—or $n$th—subinterval.

3. Cover the region with rectangles and then approximate the area under the graph of $f$ over $[a,b]$ by summing the areas of these rectangles.

a. Sketch a few rectangles. On your diagram in part 1, sketch the first three rectangles, the $i$th rectangle, and the $n$th rectangle, using the value $f$ at the right end-point of each subinterval to determine each rectangle's height.

**b.** Find a formula for the height of each rectangle. Express the height in terms of $f$ and the right end-point of the subinterval.

(1) Express the height of the first rectangle in terms of $f$ and $x_1$.

(2) Express the height of the second rectangle in terms of $f$ and $x_2$.

(3) Express the height of the $i$th rectangle in terms of $f$ and $x_i$.

(4) Express the height of the $n$th rectangle in terms of $f$ and $x_n$.

**c.** Find a formula for the area of each rectangle. Recall that the width of each rectangle is $\Delta x$.

(1) Find a formula for the area of the first rectangle.

(2) Find a formula for the area of the second rectangle.

(3) Find a formula for the area of the $i$th rectangle.

(4) Find a formula for the area of the $n$th rectangle.

**4.** Approximate the area of the region by summing the areas of the rectangles. Express the sum using sigma notation. This sum is called a *Riemann sum*.

**5.** What happens to the value of the Riemann sum if the number of rectangles is allowed to increase indefinitely? How might you express this using limit notation?

Congratulations! You have just developed the notation for what mathematicians call the *definite integral,* another extremely important calculus concept.

In the last task, you took a concrete idea—adding up the areas of rectangles—and represented it in an abstract way using mathematical notation. In particular, you considered a continuous, non-negative function $f$ defined over a closed interval $[a,b]$. You let

$$\Delta x = \frac{b - a}{n} \qquad \text{and} \qquad x_i = a + i\Delta x \quad \text{for } 0 \le i \le n$$

where $n$ is a positive integer.

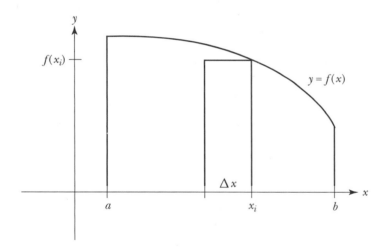

You discovered that

$$\lim_{n \to \infty} \sum_{i=1}^{n} \underbrace{f(x_i)\Delta x}$$

area of $i$th
rectangle

approximate area of
region under curve

exact area of region under the
graph of $f$ over the interval $[a,b]$

The approximate area is represented by a Riemann sum. In Task 5-9, you calculated a Riemann sum by hand. Obviously, when you do a calculation by hand, you need to keep the number of rectangles—that is, value of $n$—very small. Larger values of $n$, however, give more precise approximations. This sounds like a perfect task to do on a calculator or with the computer. In the next task, you will use ISETL to calculate some Riemann sums and observe the limiting behavior of the Riemann sums as the size of $n$ increases. Using ISETL will not only help you do the calculations, it should also help you develop a better understanding of the mathematical notation.

## Task 5-11: Calculating Riemann Sums

1. Approximate the area of the region bounded by the x-axis and the graph of

$$f(x) = x^2 - 6x + 10, \quad \text{where } 2 \le x \le 5$$

**a.** Consider the following graph of $f$.

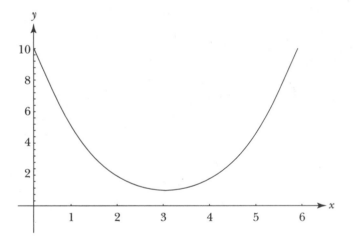

(1) Shade the area of the region under the graph of $f$ and over the closed interval [2,5].

(2) Suppose you approximate the area of this region using six rectangles, which all have the same width and where the height of each rectangle is determined by the value of $f$ at the right endpoint of its base. Draw the six rectangles, using a pencil and a straightedge.

(3) If you use six rectangles to estimate the area, will the approximation be greater than, equal to, or less than the exact value? Justify your response.

(4) Suppose you use 10 rectangles, or 100, or 1000 to estimate the area. How will the approximate value compare to the exact value?

**b.** Use ISETL to approximate the area of the region under the graph of $f(x) = x^2 - 6x + 10$ and over the closed interval [2,5], using six rectangles.

(1) Enter the following code in ISETL. Try to connect the ISETL syntax with the associated mathematical notation which you developed in Task 5-10, noting that in this example, $a = 2$, $b = 5$, and $n = 6$.

---

ISETL Code

$ Define the function $f(x) = x^2 - 6x + 10$.

```
f := func(x);
        return x**2 - 6*x + 10;
end func;
```

$ Assign values to a, b, n.

```
a := 2;
b := 5;
n := 6;
```

$ Define $\Delta x$.

```
delta_x := (b - a)/n;
```

$ Define the right end-points of the rectangles $x_i = a + i\Delta x$ where $1 \leq i \leq n$.

```
x := [a + i * delta_x : i in [1..n]];
```

$ *Note*: x is a sequence containing *n* values, whose first item x(1) corresponds to
$ $x_1$, second item x(2) corresponds to $x_2$, and so on. To display the entire sequence,
$ enter x;. To access a particular item, for instance the fifth one, enter x(5);.

$ Calculate the Riemann sum $\sum\limits_{i=1}^{n} f(x_i)\Delta x$. (See HW5.9.)

```
%+ [f(x(i)) * delta_x : i in [1..n]];
```

$ Construct a table containing the area of each of the *n* rectangles.

```
for i in [1..n] do
        writeln i, x(i), f(x(i)), f(x(i)) * delta_x;
end for;
```

$ *Note*: This table is similar to the one you constructed by hand in Task 5-9, part 4.

---

(2) Print a copy of the ISETL table and place it in the space below. Label the columns. Explain how the information in the table relates to the diagram you drew in part a.

(3) Based on your ISETL session, what is the approximate area of the region when you use a rectangular approach with the right end-point of each subinterval, where $n = 6$?

c. Use ISETL to make your approximation more precise.

*Note: To calculate the Riemann sum for a new value of n:*

i. Edit and reexecute the assignment statement for $n$.

ii. Reexecute the assignment statements for **delta_x** and **x** and the expression for the Riemann sum, which all depend on the value of $n$.

(1) Use ISETL to fill in the following chart.

| $n$ | $\sum_{i=1}^{n} f(x_i)\Delta x$ |
|-----|-----|
| 5 | |
| 15 | |
| 25 | |
| 40 | |
| 75 | |
| 100 | |

(2) As $n$ gets larger and larger, the Riemann sum gets closer and closer to the exact value of the area. Based on the chart, what might you conclude about the limiting behavior of the associated Riemann sum?

$$\sum_{i=1}^{n} f(x_i)\Delta x \rightarrow \boxed{\phantom{xx}} \quad \text{as } n \rightarrow \infty$$

2. Compare the mathematical notation developed in Task 5-10 to the associated ISETL syntax by filling in the boxes in the table given below.

*Note: A series of three dots (...) indicates that if you were considering a specific situation, you would insert more information here. For example, you would define f in the first row. Do not fill in the ...'s. Think in terms of a general situation.*

| Interpretation | Math Notation | ISETL Syntax |
|---|---|---|
| Function definition | $f(x) = \ldots$ | f := func(x);<br>end func; |
| Left end-point of interval | $a = \ldots$ | |
| Right end-point of interval | | b := ...; |
| # of rectangles | | |
| Width of $i$th rectangle | | delta_x := (b − a)/n; |
| Right end-points of $i$th subinterval | $x_i = a + i\Delta x,\ 1 \le i \le n$ | |
| Riemann Sum with $n$ rectangles | | %+[f(x(i)) * delta_x : i in [1..n]]; |

3. Consider the region bounded by the graph of $g(x) = e^x$ and the $x$-axis, where $-1 \le x \le 1$.

   a. Use your CAS to plot the region.

   (1) Place a copy of the graph in the space below.

   (2) On the graph, shade the region under consideration.

   (3) Suppose $n = 8$. Sketch the eight rectangles you would use to approximate the area of the region using the rectangular approach with the right end-point of each subinterval.

   b. Use ISETL to calculate the value of the Riemann sum for $n = 8$. Enter your result below.

   c. Explain why using the right end-point of each subinterval to determine the height of the associated rectangle will always result in an approximation that is larger than the exact value.

**d.** The area of the region under the graph of $g(x) = e^x$ and over the closed interval $-1 \le x \le 1$ is very close to 2.3504. Use ISETL to determine the minimum number of rectangles you need to consider so that the difference between the Riemann sum and 2.3504 is less than or equal to 0.05. Record the results of your trials in the table below.

| Trial Number | $n$ | $\sum_{i=1}^{n} f(x_i)\Delta x$ |
|---|---|---|
| 1 | | |
| 2 | | |
| 3 | | |
| | | |

If the limit of the Riemann sum $\sum_{i=1}^{n} f(x_i)\Delta x$ exists as $n \to \infty$, $f$ is said to be *integrable* over the closed interval $[a,b]$. A shortcut for denoting this limit is to use the definite integral notation

$$\int_a^b f(x)\ dx$$

which is read, "the integral from $a$ to $b$ of $f(x)\ dx$."

You can express a definite integral two ways: using the expression representing a function or using the name of a function. For example, in the last task, you considered the region bounded by

$$f(x) = x^2 - 6x + 10, \quad \text{where } 2 \le x \le 5$$

In this case, because you know the expression, $x^2 - 6x + 10$, as well as the name, $f$, you can represent the area under the graph of $f$ over the closed interval [2,5] by

$$\int_2^5 (x^2 - 6x + 10)\ dx \quad \text{or} \quad \int_2^5 f(x)\ dx$$

Evaluating a definite integral gives a number, such as 2.3504, $-260.5$, or 0. For instance, because the Riemann sums approximating the area under the graph of $f(x) = x^2 - 6x + 10$ over the interval [2,5] approach 6 as $n \to \infty$, you write

$$\int_2^5 (x^2 - 6x + 10)\, dx = 6$$

If the region under the graph of a function and over a closed interval has a regular shape, you can find the exact value of the associated definite integral by using the area formula for the regular shape. However, based on the work you have done so far, if the region is oddly shaped, you can only approximate the value of the definite integral by calculating a Riemann sum. The big question, of course, is how you might evaluate a definite integral without taking the limit of a Riemann sum.

The next task will help you to get used to using integral notation and to think about interpreting the notation as an area. You will convert back and forth between the description of a region and its associated definite integral. You will use a geometric approach to evaluate definite integrals that correspond to regularly shaped regions.

## Task 5-12: Interpreting Definite Integrals

1. Consider the region bounded by $f(x) = x - 4$ over the closed interval [6,10].

   **a.** Sketch the function and shade the designated region.

   **b.** Represent the area of the region by a definite integral.

   **c.** Find the value of the definite integral by using a geometric formula to calculate the area of the region.

2. Represent the area of each of the following regions by a definite integral.

   **a.** The region under the graph of $g(x) = x^4 + 3x + 8$, where $-5 \le x \le 10$.

**b.** The region bounded by $x = -2$, $x = 5$, $y = 0$, and $y = 8$.

**c.** The shaded region given below.

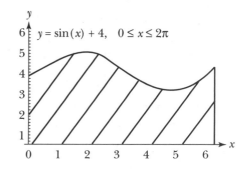

**d.** The shaded region given below.

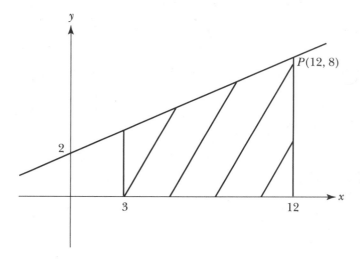

**3.** Sketch the region whose area is represented by each of the following definite integrals.

**a.** $\displaystyle\int_{-2}^{3} x^2 \, dx$

**b.** $\displaystyle\int_{-\frac{\pi}{2}}^{\frac{\pi}{2}} \cos(t) \, dt$

c. $\int_{-2}^{0} (x + 3)\, dx + \int_{0}^{2} (-x + 3)\, dx$

4. Use a geometric approach to evaluate the following definite integrals.

   *Note: Before evaluating the integral, sketch the associated region.*

   a. Evaluate $\int_{0}^{6} x\, dx.$

   b. Evaluate $\int_{-1}^{2} (2x + 6)\, dx.$

      *Note: The area of a trapezoid is given by $\frac{1}{2}(h_1 + h_2)b$, where $h_1$ and $h_2$ are the heights of the two sides and b is the width of the base.*

   c. Evaluate $\int_{-2}^{3.5} dx.$

   d. Evaluate $\int_{-3}^{1} |x|\, dx.$

   e. Evaluate $\int_{1}^{7} g(x)\, dx$, where $g$ is a semicircle with radius 3 and center at $C(4,0)$.

**f.**  Evaluate $\displaystyle\int_{-10}^{-5} 3.5\,dx.$

**5.** Suppose $f$ is a non-negative, continuous function, defined at $x = a$. Use an area-based argument to explain why $\displaystyle\int_{a}^{a} f(x)\,dx$ is equal to 0.

In this unit, you have developed definitions for the two most important concepts in calculus: the derivative and the definite integral. At first glance, these concepts seem totally unrelated—one measures rates of change, whereas the other measures the size of a quantity. It turns out that they are related and that this relationship will provide you with a straightforward way (which does not involve calculating a limit) to evaluate definite integrals. In the next task, you will take a first look at the connection between these two fundamental ideas. Discovering how this connection can help you evaluate definite integrals will have to wait until Unit 7.

Before investigating the connection, think about how integrals can be used to measure quantities that accumulate. Consider, for example, a non-negative, continuous function $f$ defined for $1 \le x \le 10$. You can find the area of the region under the graph of $f$ and over the closed interval [1,10] by covering the region with rectangles, calculating the corresponding Riemann sum, and then taking the limit. You can express this result in terms of a definite integral:

$$\int_{1}^{10} f(t)\,dt = \text{area under the graph of } f \text{ and over the closed interval } [1,10]$$

This definite integral gives the value of the area of the entire region. If, however, you only wanted to calculate the area of the region under the graph of $f$ and over the closed interval [1,6.5], you could *truncate* the Riemann sum you used to calculate the entire area—that is, delete the rectangles corresponding to [6.5,10]—and then take the limit of this *partial Riemann sum.* You can express this new result as a definite integral:

$$\int_{1}^{6.5} f(t)\,dt = \text{area under the graph of } f \text{ and over the closed interval } [1,6.5]$$

In fact, you can find the area under $f$ and over the closed interval [1,$x$], where $x$ is *any* value between 1 and 10, by taking the limit of the corresponding partial Riemann sum. As above, you can express this result as a definite integral:

$$\int_{1}^{x} f(t)\,dt = \text{area under the graph of } f \text{ and over the closed interval}$$
$$[1,x], \quad \text{where } 1 \le x \le 10$$

Since for each value of $x$ between 1 and 10 the value of the definite integral gives the area of the corresponding region, this process defines a new function, which we will call A:

$$A(x) = \int_1^x f(t)\, dt, \quad \text{where } 1 \le x \le 10$$

A is said to be the *accumulation function defined by f*, or in this case, the *cumulative area function*, because whenever you give A an $x$-value between 1 and 10, A gives the accumulated area under the graph of $f$ from 1 to $x$.

Now comes the connection between derivatives and definite integrals. According to the *Fundamental Theorem of Calculus*, the derivative of the accumulation function is equal to its underlying function, or $A'(x) = f(x)$. Consequently, you integrate $f$ to define A and differentiate A to get $f$. In the next task, you will show that this connection holds for a constant function.

## Task 5-13: Checking the Connection Between Derivatives and Definite Integrals

Consider the function $f(x) = 5$, where $2 \le x \le 9$. Define the associated cumulative area function and show that its derivative equals 5.

**1.** Sketch a graph of $f$ for $2 \le x \le 9$.

**2.** Find the cumulative area function A defined by $f$.

Note that in this case, $A(x)$ equals the area of the region under the graph of $f$ and over the closed interval $[2,x]$, where $x$ is any value between 2 and 9 inclusive.

**a.** Interpret some values of $A(x)$.

**(1)** For each of the following, make a small sketch of $f$ and shade the associated region.

**(a)** $A(4)$

**(b)** $A(6.5)$

(c)  $A(9)$

(2) Explain why $A(2) = 0$.

(3) Give the domain of $A$.

b. Use the graph of $f$ to calculate some numerical values of $A(x)$. Represent each value as a definite integral.

| | $x$ | | | | |
|---|---|---|---|---|---|
| | 2 | 3 | 7.5 | 8 | 9 |
| $A(x)$ (numerical value) | | | | | |
| $A(x)$ (represented by a definite integral) | | | | | |

c. Sketch a graph of $A$ for $2 \le x \le 9$.

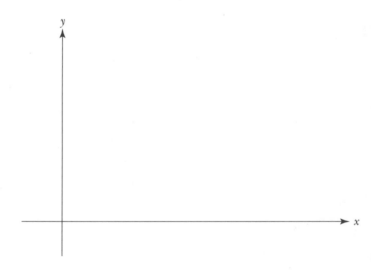

d. Represent $A$ by an expression.

3. You now have *two* functions, the original function $f(x) = 5$ and a new function $A(x) = 5x - 10$. $A$ is the accumulation function associated with $f$, which is defined by integrating $f$. The Fundamental Theorem of

Calculus claims that the derivative of the accumulation function equals the original function—that is, $A'(x) = f(x)$. So, in this instance, $A'(x)$ should equal 5. Show that this is true, using both a geometric approach and the definition of derivative.

**a.** Use a graphic approach to find $A'(x)$.

Consider the graph of $A(x) = 5x-10$ in part 2c. Show that the slope of the tangent line to any point on the graph of $A$ is 5.

**b.** Use the definition of derivative to find $A'(x)$.

Show that

$$\lim_{h \to 0} \frac{A(x + h) - A(x)}{h} = 5, \quad \text{when } A(x) = 5x - 10.$$

## Unit 5 Homework After Section 2

- Complete the tasks in Section 2. Be prepared to discuss them in class.

- In Task 5-11 you estimated the area of a given region using a rectangular approach, where the height of each rectangle was determined by the value of the function at the right end-point of the associated subinterval. Consider three other approaches for estimating this area in HW5.10.

**HW5.10** Consider, once again, the region bounded by the $x$-axis and the graph of

$$f(x) = x^2 - 6x + 10, \quad \text{where } 2 \le x \le 5$$

**1.** In parts a–c below, estimate the area of the region under the graph of $f$ and over the closed intervals [2,5] using the indicated approach. In each case:

   **i.** Sketch the graph of $f$ for $2 \le x \le 5$.

  **ii.** Subdivide the interval [2,5] into six equal subintervals. Cover the region with the specified shapes. For example, in part a, cover the region with rectangles formed by using the value of $f$ at the left end-point of each subinterval.

 **iii.** Approximate the area of the region by finding the areas of the specified shapes and summing the results.

  **a.** Rectangular approach using left end-point: Form rectangles where the height of each rectangle is determined by the value of $f$ at the left end-point of each subinterval.

  **b.** Midpoint approach: Form rectangles where the height of each rectangle is determined by the value of $f$ at the midpoint of each subinterval.

   **c.** Trapezoidal approach: Form trapezoids where the heights of each trapezoid is determined by the value of $f$ at the left and right end-points of each subinterval.

2. Which approach (rectangular approach using the right end-point, rectangular approach using the left end-point, midpoint approach, or trapezoidal approach) appears to give the most accurate result for this problem. Why?

- Investigate how the value of an approximation relates to the exact value when using different approaches.

**HW5.11** Suppose you want to approximate the area of a region under a given curve and over a closed interval. For each of the following statements:

   **i.** Sketch a graph of a function for which the statement is true.
   **ii.** Shade the region.
   **iii.** Sketch a typical rectangle or trapezoid. (Your example should work for any number of rectangles or trapezoids.)

1. You approximate the area using the rectangular approach with the left end-point of each subinterval. The approximate area is greater than the exact area.

2. You approximate the area using the rectangular approach with the left end-point of each subinterval. The approximate area is less than the exact area.

3. You approximate the area using the rectangular approach with the left end-point of each subinterval. The approximate area equals the exact area.

4. You approximate the area using the rectangular approach with the right end-point of each subinterval. The approximate area is greater than the exact area.

5. You approximate the area using the rectangular approach with the right end-point of each subinterval. The approximate area is less than the exact area.

6. You approximate the area using the rectangular approach with the right end-point of each subinterval. The approximate area equals the exact area.

7. You approximate the area using the trapezoidal approach. The approximate area is greater than the exact area.

8. You approximate the area using the trapezoidal approach. The approximate area is less than the exact area.

9. You approximate the area using the trapezoidal approach. The approximate area equals the exact area.

- Model some situations and interpret some definite integrals in HW5.12.

**HW5.12**

1. Model some familiar situations and represent the regions by definite integrals.

Look again at the dock and the boat house which you considered in Task 5-7, parts 1 and 4. In each case:

   **i.** Model the situation by placing the region on a coordinate system and finding an expression for the function which defines the shape of the region.

   **ii.** Express the exact area of the region in terms of a definite integral using the expression from part i.

   **iii.** Evaluate the integral using a geometric approach.

   **a.** The deck of the dock.

   **b.** The back of the boat house. (*Note:* The upper boundary of the back of the boat house is modeled by two expressions. Consequently, its area can be represented by the sum of two definite integrals. It is also possible to represent the area as a multiple of a single definite integral. Try to do it both ways.)

**2.** Consider

$$h(r) = \begin{cases} r + 1, & \text{if } r \le 3 \\ 4, & \text{if } 3 < r < 5 \\ -4r + 24, & \text{if } r \ge 5 \end{cases}$$

Use a geometric approach to evaluate $\int_0^6 h(r)\,dr$.

**3.** Sketch the region whose area is given by $\int_0^2 x^2\,dx + \int_2^4 (x-4)^2\,dx$.

**4.** Suppose $f$ is a non-negative continuous function with domain $1 \le x \le 3$, where the area of the region under the graph of $f$ over the closed interval $[1,3]$ is greater than or equal to $\frac{1}{2}f(1.5) + \frac{1}{2}f(2) + \frac{1}{2}f(2.5) + \frac{1}{2}f(3)$. Sketch a graph of $f$.

**5.** Sketch the graph of a function $g$, where $\int_{-4}^{-1} g(x)\,dx = 15/2$.

**6.** Explain why the following inequalities are true.

   **a.** $\int_0^\pi \sin(\theta)\,d\theta < \pi$

   **b.** $\int_0^6 f(t)\,dt < 24$, where

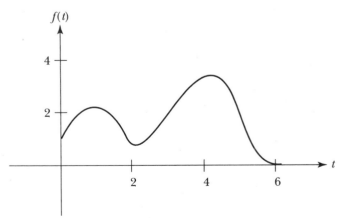

- Show that the Fundamental Theorem of Calculus holds for a linear function.

**HW5.13** Consider $f(x) = 2x$, where $0 \leq x \leq 4$.

1. Graph $f$ for $0 \leq x \leq 4$.

2. Let $A$ be the accumulation function defined by $f$.

   **a.** Interpret some values of $A(x)$. For each of the following values of $x$:

   **i.** Draw a small sketch of $f$ and shade the region whose area equals $A(x)$.
   **ii.** Find the numerical value of $A(x)$.
   **iii.** Represent $A(x)$ by a definite integral.

   **(1)** $x = 0$      **(4)** $x = 2.5$
   **(2)** $x = 1$      **(5)** $x = 3$
   **(3)** $x = 2$      **(6)** $x = 4$

   **b.** Sketch a graph of $A$ for $0 \leq x \leq 4$.

   **c.** Find an expression that fits the graph of $A$.

3. Show that the Fundamental Theorem of Calculus holds. Use the definition of the derivative to show that $A'(x) = f(x)$.

- Describe your understanding of definite integral in HW5.14.

**HW5.14** What is a definite integral? Write a short essay explaining what a definite integral is. In your essay:

   **i.** Give the definition of a definite integral as the limit of a Riemann sum.
   **ii.** Support your definition by sketching and labeling an appropriate diagram.

- Write your journal entry for this unit. As usual, before you begin to write, review the material in the unit. Think about how it all fits together. Try to identify what, if anything, is still causing you trouble.

**HW5.15** Write your journal entry for Unit 5.

1. Reflect on what you have learned in this unit. Describe in your own words the concepts that you studied and what you learned about them. How do they fit together? What concepts were easy? Hard? What were the important ideas? Give some examples of the main ideas.

2. Reflect on the learning environment for the course. Describe the aspects of this unit and the learning environment that helped you understand the concepts you studied. What activities did you like? Dislike?

# *Unit 6:*

# DERIVATIVES: THE CALCULUS APPROACH

*In symbols one observes an advantage in discovery which is greatest when they express the exact nature of a thing briefly and, as it were, picture it; then indeed the labor of thought is wonderfully diminished.*

Gottfried Wilhelm Leibniz (1646–1716), one of the greatest inventors of mathematical symbols, in a letter to his friend Tschirnhaus. From George F. Simmons, *Calculus Gems: Brief Lives of Memorable Mathematicians*, McGraw-Hill, 1992.

## OBJECTIVES

**1.** Find derivatives of constant, power, trigonometric, exponential, and logarithmic functions.

**2.** Develop rules for finding the derivative of a combination of functions.

**3.** Use derivatives to analyze functional behavior.

## OVERVIEW

In Unit 5, you developed the mathematical notation for finding the slope of the tangent line when $x = a$ for a smooth function $f$. In the process, you discovered a definition for one of the most important concepts in calculus: the *derivative* of a function.

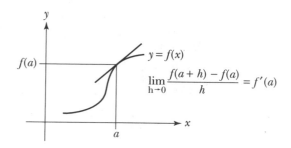

You used the definition to evaluate derivatives. You observed that the derivative of a function is itself a function, whose domain is the set of values in the domain of the underlying function where the tangent lines to the graph exist and whose output for a given input value equals the slope of the associated tangent line. You explored the relationship between a given function and its derivative, observing how the tangent line behaves as it travels along the curve and asking questions such as: When is the derivative zero? When is the derivative positive/negative? What does this tell you about the behavior of the slope of the tangent line to the curve? What does this tell you about the shape of the graph of the function? In the process, you developed a statement of the First Derivative Test.

At this point, you should have a solid conceptual understanding of what a derivative is and what it tells you about the behavior of the underlying function. To get information about a function from its derivative, you need to be able to find the derivative. Unfortunately, using the limit-based definition to do so may involve doing some messy algebraic computations. There has to be an easier way to do this; and there is. Discovering some straightforward rules for finding derivatives is what this unit is all about.

Using the definition to find the derivative of a function at $x = a$, however, does help you keep in mind what the derivative represents. For instance, the derivative represents *the slope of the tangent line* to the graph of $f$ at $P(a,f(a))$. This follows since,

**i.** The *difference quotient* equals the slope of the secant line determined by the points $P(a,f(a))$ and $Q(a + h,f(a + h))$. That is,

$$\text{slope} = \frac{f(a + h) - f(a)}{(a + h) - a} = \frac{f(a + h) - f(a)}{h}.$$

**ii.** As $h \to 0$, $a + h$ gets closer and closer to $a$, and as a result, the slope of the secant line determined by $P$ and $Q$ gets closer and closer to the slope of the tangent line to the graph of $f$ at $P$.

In addition to representing the slope of the tangent line, the derivative represents *the (instantaneous) rate of change* of the function $f$ at $x = a$. This interpretation follows from the following facts:

**i.** The difference quotient equals the average rate of change of $f$ from $x = a$ to $x = a + h$. That is,

$$\frac{\text{change in the output values}}{\text{change in the input values}} = \frac{f(a + h) - f(a)}{(a + h) - a} = \frac{f(a + h) - f(a)}{h}$$

**ii.** As $h \to 0$, $a + h$ gets closer and closer to $a$, and as a result the average rate of change gets closer and closer to the (instantaneous) rate of change of $f$ at $x = a$.

As you use the rules, try not to lose sight of what a derivative represents and how it can be constructed.

The rules provide techniques for finding derivatives of basic functions, including:

- Constant functions

$$f(x) = 6.75$$
$$g(t) = -\pi$$

- Power functions

$$r(x) = x^{10}$$
$$s(w) = w^{-1/3}$$

- Trigonometric functions

$$f(\theta) = \sin(\theta)$$
$$g(\theta) = \cos(\theta)$$

- Exponential and logarithmic functions

$$y = e^x$$
$$f(x) = 2^x$$
$$y = \ln(x)$$
$$p(t) = \log_2(t)$$

The rules also provide techniques for finding derivatives of combinations of functions, including:

- The scalar multiple of a function

$$f(t) = 4(t^{1/2} + 2)$$
$$n(x) = -5\sin(x^2)$$

- The sum or difference of two functions

$$p(r) = e^{2r^2} + e^{(-r)}$$
$$q(x) = \sqrt{x} - \frac{2}{x}$$

- The product of two functions

$$f(x) = \sqrt{x}\ln(1 + x)$$
$$g(t) = (t^3 - 2t^2 + 5)(t^{10} - 4)$$

- The quotient of two functions

$$s(x) = \frac{x^2 - 3}{2x + 1}$$

$$t(\theta) = \frac{\sin(\theta + 2)}{\cos(\theta^2)}$$

- The composition of two functions

$$k(r) = (r^3 - 1)^{10}$$
$$f(t) = \ln(\sqrt{t^2 - 1})$$
$$s(\theta) = \sin(\cos(\theta))$$

Finding the derivative of a complicated function usually involves using more than one rule. In order to use the rules, you will need to "uncombine" the function and identify its basic parts.

Sharpen your pencils. You will be finding most of the derivatives by hand. As you probably suspect, a computer algebra system can do most of the calculations that you will be doing. But you need to learn how to differentiate with pencil and paper before you turn to your CAS. Understanding the process of differentiation will enable you to think about how the computer might be processing the information that you give it and properly interpret its output.

---

## SECTION 1

## Differentiating Combinations of Functions

The differentiation rules can be proved by applying the definition of derivative to an expression representing the general function. For example, in Unit 5, you observed that the slope of every tangent line to the graph of a constant function is 0, since the graph of a constant function is always a horizontal line, and then in HW5.1, you used the definition of derivative to prove that this is the case. You considered the general function $f(x) = c$, where $c$ is any constant, and used the definition of derivative to show that $f'(x) = 0$. This result is your first differentiation rule, which is called the *Constant Rule.*

---

**Constant Rule**

Let $c$ be a constant.

- If $f(x) = c$, then $f'(x) = 0$.

In other words,

- "The derivative of a constant function is 0."

---

According to the Constant Rule,

$$\text{if } f(x) = 6.75, \text{ then } f'(x) = 0,$$

or

$$\text{if } g(t) = -\pi, \text{ then } g'(t) = 0.$$

Since $f$ and $g$ are constant functions, the rule says their derivatives are 0, and that's all there is to it.

In the following tasks, you will examine rules for finding the derivative of various combinations of functions. In some cases, you will use the definition of derivative to prove a rule. In other cases, you will convince yourself that a rule holds by considering some specific examples and then generalizing a pattern which emerges, by using your conceptual understanding of the meaning of a derivative, or by showing that a new rule gives the same result as a combination of rules you already know.

The first rule you will examine is called the *Power Rule.*

---

**Power Rule**

Let $n$ be a real number.

- If $f(x) = x^n$, then $f'(x) = nx^{n-1}$.

- "The derivative of $x$ raised to the $n$th power equals $n$ times $x$ raised to the $(n-1)$st power."

---

A few comments before you begin. First, a comment about notation. As you know, frequently there is more than one way to say something, and this is certainly the case when referring to the derivative of a function. Until now, you have used $f'$ to denote the general derivative of the function $f$. For example, if $f(x) = x^5$, then according to the Power Rule

$$f'(x) = 5x^{5-1} = 5x^4.$$

In addition to using the "prime" notation for a derivative, you can use $\dfrac{d}{dx}$ to indicate the general derivative with respect to $x$. For example, if $y = x^5$, you can write

$$\frac{dy}{dx} = \frac{d}{dx}(x^5) = 5x^4.$$

Consequently, another way to state the Power Rule is

$$\frac{d}{dx}(x^n) = nx^{n-1}.$$

In the future, each new rule will be stated three ways: using the prime notation, the $\dfrac{d}{dx}$ notation, and a verbal description.

Second, a comment about names. Although all the rules will be stated in terms of $f$ and $x$, there is nothing special about these names: $f$ is a common name for a function and $x$ is a common name for the independent variable of a function. You can, of course, give a function or a variable any name you want. Usually, the names reflect the situation they represent. For instance, in the case of a function of time, the independent variable is frequently named $t$, whereas in the case of a trigonometric function, the independent variable is often called $\theta$.

Finally, a comment about terminology. When we refer to a derivative, we mean *the derivative with respect to the independent variable.* Therefore, if $f$ is a function of $x$, then $f'$ is also a function of $x$, and it is referred to as "the derivative of $f$ with respect to $x$." In this case, the derivative is represented by

$$f'(x) \quad \text{or} \quad \frac{d}{dx}(f(x)).$$

Similarly, if $g$ is a function of $\theta$, $g'$ is "the derivative of $g$ with respect to $\theta$," and it is represented by

$$g'(\theta) \quad \text{or} \quad \frac{d}{d\theta}(g(\theta)).$$

---

## Task 6-1: Examining the Power Rule

*In HW5.1, you showed that the Power Rule holds when* n = 1 *and when* n = 3.

- *In the case of* n = 1, *you observed that the slope of the tangent line to the graph of* f(x) = mx + b *is always* m, *and as a result,* f′(x) = m. *For example, if* f(x) = x, *then* f′(x) = 1. *The Power Rule gives the same outcome.*

$$f(x) = x^1,$$
$$f'(x) = 1x^{1-1} = x^0 = 1.$$

- *In the case of* n = 3, *you used the definition of derivative to prove that if* f(x) = x³, *then* f′(x) = 3x². *Again, the Power Rule leads to the same result.*

$$f(x) = x^3,$$
$$f'(x) = 3x^{3-1} = 3x^2.$$

1. Show that the Power Rule holds when $n = 2$.

   a. Use the Power Rule to find the derivative of $f(x) = x^2$.

**b.** Use the definition of derivative, namely

$$f'(x) = \lim_{h \to 0} \frac{f(x+h) - f(x)}{h}$$

to find the derivative of $f(x) = x^2$.

**c.** Compare the results in parts a and b. They should be the same. If they are not, check your work.

**2.** Not only does the Power Rule hold when $n = 1, 2,$ and $3$, it also holds when $n$ is any real number (see HW6.6).

Use the rule to calculate the derivative of the following functions. Be sure and show all your work.

*Note: Before applying the* **Power Rule***, you may have to rewrite the expression representing the function as x raised to a power. For example,*
$f(x) = \sqrt[5]{x^4} = x^{4/5}$.

**a.** $f(x) = x^{10}$

$f'(x) =$

**b.** $y = x^{-1/3}$

$\dfrac{dy}{dx} =$

**c.** $g(t) = \sqrt{t}$

$g'(t) =$

**d.** $h(r) = \dfrac{1}{r^2}$

$h'(r) =$

**e.** $w = \dfrac{1}{\sqrt[4]{t}}$

$\dfrac{dw}{dt} =$

**3.** Use the Power Rule to calculate the following values.

  **a.** Consider $f(x) = x^4$.

   **(1)** Find the slope of the tangent line to the graph of $f$ when $x = 3$.

   *Hint: First find an expression for $f'(x)$. Then evaluate the expression at $x = 3$—that is, find $f'(3)$.*

   **(2)** Find the slope of the tangent line to the graph of $f$ at the point $P(-3, f(-3))$.

  **b.** Consider $y = \dfrac{1}{t^4}$.

   **(1)** Find the rate of change when $t = -2$.

   **(2)** Find the rate of change when $t = \dfrac{1}{2}$.

  **c.** Find the value of $g'(8)$ if $g(w) = \sqrt[3]{w^2}$.

At this point, you can differentiate two basic functions: constant functions and power functions. Next, you will find your first rule for a combination of functions, namely a scalar multiple of a function. Suppose $f$ is a given function and $c$ is fixed constant. Define a new function $k$, where $k(x) = c\, f(x)$. The following sequence of steps shows how the definition of derivative can be used to find $k'$.

$$k'(x) = \lim_{h \to 0} \frac{k(x + h) - k(x)}{h}$$

$$= \lim_{h \to 0} \frac{c\, f(x + h) - c\, f(x)}{h}$$

$$= \lim_{h \to 0} c\left( \frac{f(x + h) - f(x)}{h} \right)$$

$$= c \lim_{h \to 0} \frac{f(x + h) - f(x)}{h}$$

$$k'(x) = c\, f'(x).$$

This gives the *Scalar Multiple Rule*:

---

**Scalar Multiple Rule**

Let $f$ be a given function and $c$ be a fixed constant.

- If $k(x) = c f(x)$, then $k'(x) = c f'(x)$.
- $\dfrac{d}{dx}(c f(x)) = c \dfrac{d}{dx}(f(x))$.
- "The derivative of a constant times a function equals the constant times the derivative of the function."

---

## Task 6-2: Applying the Scalar Multiple Rule

1. Assume $k(x) = c f(x)$, where $f$ is a given function and $c$ is fixed constant. Provide an explanation for each step in the proof of the Scalar Multiple of a Function Rule, which is given above. To help you get started, the explanation for the first step is provided.

   *Note: One way to "prove" a rule is to start with one side of the rule—in this case, $k'(x)$—and use facts you know, such as the definition of derivative, to get to the other side of the rule—in this case, $c f'(x)$. Try this.*

   | Step | Explanation |
   |---|---|
   | $k'(x) = \lim\limits_{h \to 0} \dfrac{k(x + h) - k(x)}{h}$ | Apply the definition of derivative to $k$. |

   $$= \lim_{h \to 0} \frac{c f(x + h) - c f(x)}{h}$$

   $$= \lim_{h \to 0} c \left( \frac{f(x + h) - f(x)}{h} \right)$$

   $$= c \lim_{h \to 0} \frac{f(x + h) - f(x)}{h}$$

   $$k'(x) = c f'(x)$$

2. Use the Scalar Multiple Rule in conjunction with the Power Rule to find the derivative of each of the following functions. As always, show your work.

   **a.** $y = 10.5x^3$

   $$\frac{dy}{dx} =$$

   **b.** $f(x) = -4x^{-2}$

   $$f'(x) =$$

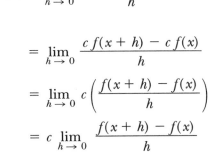

c. $g(s) = \frac{1}{2}s^{2/3}$

$g'(s) =$

d. $p = \dfrac{5}{t^3}$

$\dfrac{dp}{dt} =$

e. $f(x) = -\dfrac{8}{3x^{1/8}}$

$f'(x) =$

3. Consider $f(x) = 3x^2$. For each of the values listed below:
   (i) Find the slope of the tangent line to the graph of $f$ when $x$ has the given value.
   (ii) Indicate if $f$ is increasing, decreasing, or has a possible turning point when $x$ has the given value.
   a. $x = -2.5$

   b. $x = 0$

   c. $x = 2.5$

4. Find the equation of the tangent line to the graph of $y = 2\sqrt{x}$ when $x = 9$.

The Scalar Multiple Rule, used in conjunction with the Power Rule, provides you with a method for finding the derivative of a constant times $x$ raised to a power, such as $5x^{10}$ or $\frac{2}{3}x^{3/2}$. The next step is to find the derivative of the sum (or difference) of several terms having this form, such as the derivative of the polynomial $y = 2x^3 - 6x^2 + 4x - 9$. The Sum and Difference Rules provide you with a way to do this.

## Task 6-3: Using the Sum and Difference Rules

**1.** The *Sum Rule* states that

---

**Sum Rule**

Let $f$ and $g$ be given functions.

- If $k(x) = (f + g)(x)$, then $k'(x) = f'(x) + g'(x)$.
- $\frac{d}{dx}(f(x) + g(x)) = \frac{d}{dx}(f(x)) + \frac{d}{dx}(g(x))$.
- "The derivative of a sum equals the sum of the derivatives."

---

Use the definition of derivative to show that the Sum Rule holds.

**a.** Before beginning the proof, recall that to evaluate the sum of two functions at an input value, you evaluate each function at the given value and then add the results. Thus, by definition,

$$(f + g)(x) = f(x) + g(x).$$

Use this definition to evaluate $f + g$ at $x + h$. That is, evaluate

$$(f + g)(x + h) = \underline{\hspace{4cm}}.$$

Note: $(f + g)(x + h)$ is not equal to $f(x) + f(h) + g(x) + g(h)$. If you thought it was, stop and figure out why it isn't before you do part b.

**b.** Prove the Sum Rule by filling in the step corresponding to each the explanations in the table below.

| Step | Explanation |
|---|---|
| $k'(x) = \lim\limits_{h \to 0} \dfrac{k(x + h) - k(x)}{h}$ | Apply the definition of derivative to $k$. |
| $=$ | Substitute $f + g$ for $k$. |
| $=$ | Evaluate $f + g$ at $x + h$ and $x$. (See part a.) |
| $=$ | Gather together the $f$ terms and the $g$ terms. |

| <u>Step</u> | <u>Explanation</u> |
|---|---|
| = | Express the quotient as the sum of the difference quotients for $f$ and $g$. |
| = | Express the limit as the sum of two limits. |
| $k'(x) = f'(x) + g'(x)$ | Use the definition of derivative (twice!). |

**2.** The *Difference Rule* states that

---

**Difference Rule**

Let $f$ and $g$ be given functions.

- If $k(x) = (f - g)(x)$, and $k'(x) = f'(x) - g'(x)$.
- $\dfrac{d}{dx}(f(x) - g(x)) = \dfrac{d}{dx}(f(x)) - \dfrac{d}{dx}(g(x))$.
- "The derivative of a difference equals the difference of the derivatives."

---

The proof of the Difference Rule is similar to the proof of the Sum Rule. You just change the plus signs to minus signs in the appropriate places.

Use the Sum and Difference Rule in conjunction with the Power and Constant Times a Function Rules to find the derivatives of the following functions.

*Note: Before finding the derivative, express each term in the function as a constant times x raised to a power. For example, before differentiating, rewrite*

$$f(x) = \frac{x^3}{2} - 6\sqrt[3]{x} + \frac{4}{x^3}$$

*as*

$$f(x) = \frac{1}{2}x^3 - 6x^{1/3} + 4x^{-3}.$$

**a.** $y = 3x^{10} - 9x^5 + 2$

**b.** $s = \dfrac{9}{t^3} + \dfrac{t^3}{9} - 9$

   **c.** $g(x) = \frac{2}{3}x^{3/2} + 2\sqrt{x} - \frac{1}{5}x^{5/2}$

   **d.** $u(x) = 6x^{-3} - \dfrac{x^{-2}}{2}$

   **e.** $r = 2\sqrt{v} - \dfrac{6}{\sqrt{v}} + 5$

**3.** Consider $f(x) = x^3 - 3x$.

   **a.** Find $f'(x)$.

   **b.** Find the equation of the tangent line to the graph of $f$ when $x = 0$.

   **c.** Find the $(x, y)$-coordinates of the points on the graph of $f$ where the tangent lines are horizontal.

   **d.** Find the $(x, y)$-coordinates of all the points on the graph of $f$ where the slope of the tangent line is 9.

The Power Rule gives you a way of finding the derivative of a single variable raised to a rational power, $x^n$. The *Extended Power Rule* provides you with a technique for finding the derivative of a function raised to a power,

$(f(x))^n$. For instance, the Extended Power Rule enables you to find the derivative of functions such as

$$g(x) = (x^4 - 6x^3)^{10} \quad \text{and} \quad h(t) = \sqrt[3]{t^4 - 6t^3} = (t^4 - 6t^3)^{1/3}$$

This new rules says:

---

**Extended Power Rule**

Let $f$ be a function and $n$ be a real number.

- If $k(x) = (f(x))^n$, then $k'(x) = n(f(x))^{n-1} f'(x)$.
- $\frac{d}{dx}((f(x))^n) = n(f(x))^{n-1} \frac{d}{dx}(f(x))$.
- "The derivative of a function raised to the $n$th power is $n$ times the function raised to the $(n-1)$st power times the derivative of the function."

---

For example, according to the Extended Power Rule:

$$\text{If } h(t) = (t^4 - 6t^3)^{1/3}, \text{ then}$$
$$h'(t) = \frac{1}{3}(t^4 - 6t^3)^{(1/3)-1} \frac{d}{dt}(t^4 - 6t^3)$$
$$= \frac{1}{3}(t^4 - 6t^3)^{-2/3}(4t^3 - 18t^2).$$

In the next task, you will try to convince yourself that the Extended Power Rule holds by doing the following:

(1) Find the derivative using the Extended Power Rule.
(2) Simplify the given expression representing the function and then find the derivative without using the new rule.
(3) Compare the results of (1) and (2), which should be the same, since the Extended Power Rule really does hold.

---

## Task 6-4: Employing the Extended Power Rule

1. Show that the Extended Power Rule holds for the following functions.

   **a.** Consider $y = (x^2)^3$.

     **(1)** Use the Extended Power Rule to find $\frac{dy}{dx}$.

     **(2)** Simplify the given expression for $y$. Then find $\frac{dy}{dx}$ (without using the Extended Power Rule).

**(3)** Show that the Extended Power Rule holds for this example by showing that your answers to (1) and (2) are equivalent.

**b.** Consider $s(t) = (t^3 - 6)^2$.

    **(1)** Use the Extended Power Rule to find $s'(t)$.

    **(2)** Expand the expression representing $s$. Then find $s'(t)$ (without using the Extended Power Rule).

    **(3)** Show that the Extended Power Rule holds for this example by showing that your answers to (1) and (2) are equivalent.

**2.** Based on the examples in part 1, it appears that the Extended Power Rule does hold. Use the Extended Power Rule, along with all the other rules you have developed, to differentiate the following functions. Show all your work.

    **a.** $y = (x^3 - 6)^{101}$

    **b.** $z = \dfrac{1}{(t^2 + 2t - 1)^2}$

    **c.** $g(r) = \left( \dfrac{1}{r} + \dfrac{1}{r^2} \right)^{10}$

**d.** $y = \sqrt[3]{p^3 + 1}$

**e.** $f(x) = \dfrac{-4}{(x^4 + 2x^2 + 4x)^{1/4}}$

**3.** Consider the function $h(x) = (x^2 - 9)^{10}$. Find all the possible values of $x$ where $h$ might have a local maximum or a local minimum.

In the next two tasks, you will examine the rules for finding the derivative of the product and the quotient of two functions where

$$(f \cdot g)(x) = f(x) \cdot g(x),$$
$$\left(\frac{f}{g}\right)(x) = \frac{f(x)}{g(x)}.$$

Start with the *Product Rule* which states:

---

**Product Rule**

Let $f$ and $g$ be given functions.

- If $k(x) = (f \cdot g)(x)$, then $k'(x) = f(x) \cdot g'(x) + g(x) \cdot f'(x)$.
- $\dfrac{d}{dx}(f(x)g(x)) = f(x)\dfrac{d}{dx}(g(x)) + g(x)\dfrac{d}{dx}(f(x))$.
- "The derivative of a product equals the first function times the derivative of the second plus the second times the derivative of the first."

---

## Task 6-5: Investigating the Product Rule

**1.** Convince yourself that the Product Rule holds, for the functions given below, by showing that the new rule gives the same result as the rules you examined previously.

**a.** $y = (x^3)(x^7)$

   **(1)** Use the Product Rule to find the derivative of $y$.

(2) Combine the $x$ terms on the right-hand side of $y$. Then find the derivative of the result without using the Product Rule.

(3) Show that the Product Rule holds in this case by showing that the results of (1) and (2) are equivalent.

**b.** $h(u) = 4u^2(u^3 - 3u)$

(1) Use the Product Rule to find the derivative.

(2) Multiply together the terms on the right-hand side of $h$. Then find the derivative of the result without using the Product Rule.

(3) Show that the Product Rule holds in this case by showing that the results of (1) and (2) are equivalent.

2. Use the Product Rule in conjunction with the other rules to calculate the derivatives of the following functions.

**a.** $y = (x^2 + 6)^4(x - 1)$

**b.** $x(t) = \sqrt{t}(2t + 4)$.

**c.** $s(n) = \dfrac{1}{n^4}(n^3 - 2n^2 + 4n - 9)^4$

**d.** $q = (\sqrt{r} - 1)(\sqrt{r} + 1)$

**e.** $f(x) = \left(\dfrac{1}{x} - 1\right)^{-6}\left(\dfrac{2}{x^2}\right)$

3. Recall that the sign chart for a function indicates the values where the function is 0 and the intervals where it is positive or negative. (See Task 5-5 and Task 5-6.)

   Consider $f(x) = x(x - 1)^2$. Find the sign chart for $f'$ as you complete parts a through c given below.

   Sign of $f'$

   $$\longrightarrow x$$

   Info re $f$

   **a.** Set $f'(x) = 0$ and solve for $x$. Indicate the location of these values by placing a 0 above the axis for the sign chart for $f'$.

   **b.** Mark $+$'s above the axis for the sign chart to indicate the intervals where $f'(x) > 0$. Mark $-$'s to indicate the intervals where $f'(x) < 0$.

   *Hint: Based on part a, you know the locations of all the zeros of f'(x). In between two adjacent zeros, f'(x) is either positive or negative; it cannot be both, otherwise f' would have another zero. Similarly, f'(x) must be either positive or negative to the left of the leftmost zero and to the right of the rightmost zero.*

*Consequently, one way to find the sign of f′ is to evaluate f′(x) at a value to the left of the leftmost zero, at a value between each pair of adjacent zeros, and at a value to the right of the rightmost zero. In each case, the sign of the result is the sign of f′(x) in that interval.*
*Use this approach to fill in +'s and −'s for the sign of f′.*

**c.** Recall that if $f'$ is positive, then the graph of $f$ is increasing since the slope of the tangent line is positive. Similarly, if $f'$ is negative, then the graph of $f$ is decreasing.

    **(1)** Fill in the Info re $f$ on the sign chart for $f'$, indicating the intervals where the graph of $f$ is increasing ("inc") and decreasing ("dec").

    **(2)** Describe the general shape of the graph of $f$.

The Quotient Rule states that

---

**Quotient Rule**

Let $f$ and $g$ be given functions.

- If $k(x) = \left(\dfrac{f}{g}\right)(x)$, then $k'(x) = \dfrac{g(x)f'(x) - f(x)g'(x)}{(g(x))^2}$.

- $\dfrac{d}{dx}\left(\dfrac{f(x)}{g(x)}\right) = \dfrac{g(x)\dfrac{d}{dx}(f(x)) - f(x)\dfrac{d}{dx}(g(x))}{(g(x))^2}$.

- "The derivative of a quotient equals the bottom (denominator) times the derivative of the top (numerator) minus the top times the derivative of the bottom all over the bottom squared."

---

The quotient of two functions can always be expressed as a product of two functions:

$$\frac{f(x)}{g(x)} = f(x)(g(x))^{-1}.$$

Consequently, the Product Rule can be used to prove the Quotient Rule. In the homework problems at the end of the section, you will have an opportunity to do this. For now, convince yourself that the Quotient Rule holds by differentiating some functions using the rules you already know and using the Quotient Rule. Then show that the results are the equivalent.

## Task 6-6: Engaging the Quotient Rule

**1.** Show that the Quotient Rule holds for the following functions.

**a.** $y = \dfrac{2x^4 - x^3}{x^2}$

**(1)** Use the Quotient Rule to find $\dfrac{dy}{dx}$.

**(2)** Now, find the derivative without using the Quotient Rule. First, simplify the expression representing $y$. Then find $\dfrac{dy}{dx}$ using the previous rules.

**(3)** Show that the Quotient Rule holds for the given function by showing that your answers to (1) and (2) are equivalent.

**b.** $q(m) = \dfrac{m}{m + 2}$

**(1)** Use the Quotient Rule to find $q'(m)$.

**(2)** Now, find the derivative without using the Quotient Rule. First, rewrite the expression representing function $q$ as a product. Then use the Product Rule to find the derivative.

(3) Show that the Quotient Rule holds for the given function by show-ing that your answers to (1) and (2) are equivalent.

2. Now that you are convinced that the Quotient Rule works, use it to dif-ferentiate the following functions.

**a.** $y = \dfrac{x^2 + 1}{6 - x^3}$

**b.** $f(t) = \dfrac{1 + \sqrt{t}}{1 - \sqrt{t}}$

**c.** $y(p) = \dfrac{p}{4p^4 + 3p^3 + 2p^2 + p}$

**d.** $w = \dfrac{b^2 + b - 9}{\sqrt{b}}$

**e.** $f(x) = \dfrac{(x^2 - 1)^3}{6 - x^4}$

The Extended Power Rule is a special case of a powerful rule called the Chain Rule, which provides a technique for finding the derivative of the composition of two functions. According to the *Chain Rule*:

---

**Chain Rule**

- If $k(x) = (g \circ f)(x)$, then $k'(x) = g'(f(x))f'(x)$.
- $\dfrac{d}{dx}(g(f(x))) = g'(f(x))f'(x)$.
- "The derivative of the composition of two functions is the derivative of the outer function evaluated at the inner function times the derivative of the inner function."

---

In the next task, you will show that the Chain Rule and the Extended Power Rule lead to the same result.

## Task 6-7: Utilizing the Chain Rule

**1.** Show that the Chain Rule holds for the following function:

$$k(x) = \sqrt[3]{x^4 - 6x^3} = (x^4 - 6x^3)^{1/3}.$$

**a.** Find the derivative of $k$ using the Extended Power Rule.

**b.** Now find the derivative of $k$ using the Chain Rule and show that the result is the same as in part a.

**(1)** First "uncombine" $k$ and express $k$ as the composition of two functions $f$ and $g$ such that

$$k(x) = (g \circ f)(x) = g(f(x)).$$

$$f(x) = \underline{\hspace{2cm}}$$

$$g(x) = \underline{\hspace{2cm}}$$

**(2)** Check that your choices for $f$ and $g$ are correct by showing that $g(f(x)) = (x^4 - 6x^3)^{1/3}$.

**(3)** Find the derivative of $f \circ g$ using the Chain Rule.

  **(a)** Find $f'(x)$.

  **(b)** Find $g'(x)$.

  **(c)** Find $g'(f(x))f'(x)$.

  *Note: This result should be equivalent to the derivative you found using the Extended Power Rule in part a. If it isn't, check your work.*

**2.** The Extended Power Rule is a "special case" of the Chain Rule. Show that this is the case.

  **a.** Represent $(f(x))^n$ as the composition of two functions $f \circ g$. What is $g(x)$ in this case?

  **b.** Use the Chain Rule to find the derivative of the composition.

**c.** Using the Chain Rule should give you the Extended Power Rule. If it doesn't, check your work.

3. For each of the following functions, express the function as the composition of two functions and then use the Chain Rule to find its derivative. Show all your work.

**a.** $h(t) = (3t^2 - 5t + 7)^{-1}$

**b.** $S(x) = \left( \dfrac{3x + 4}{6x - 7} \right)^3$

---

## Summary of the Differentiation Rules, Part 1

**Constant Rule**

$$\frac{d}{dx}(c) = 0$$

**Power Rule**

$$\frac{d}{dx}(x^n) = nx^{n-1}$$

## Scalar Multiple Rule

$$\frac{d}{dx}(c\,f(x)) = c\,f'(x)$$

## Sum and Difference Rules

$$\frac{d}{dx}(f(x) \pm g(x)) = f'(x) \pm g'(x)$$

## Extended Power Rule

$$\frac{d}{dx}((f(x))^n) = n(f(x))^{n-1}f'(x)$$

## Product Rule

$$\frac{d}{dx}(f(x)g(x)) = f(x)g'(x) + g(x)f'(x)$$

## Quotient Rule

$$\frac{d}{dx}\left(\frac{f(x)}{g(x)}\right) = \frac{g(x)f'(x) - f(x)g'(x)}{(g(x))^2}$$

## Chain Rule

$$\frac{d}{dx}(g(f(x))) = g'(f(x))f'(x)$$

## Unit 6 Homework After Section 1

- Complete the tasks in Section 1. Be prepared to discuss them in class.

- Calculate some derivatives using the rules developed in this section in HW6.1.

**HW6.1** Differentiate the following functions using pencil and paper.

**1.** $f(x) = 5(x^6 - 3x^2)^3$

**2.** $s(t) = t^3\sqrt{t^{1/5}} - 7t^3 + 13$

**3.** $y = 4\left(\dfrac{x^4 - 3x + 1}{2x^2}\right)$

**4.** $y = \dfrac{(x^2 + 6x)(4x^3 - 9x^2)}{x - 5}$

**5.** $g(r) = r^{1/2}(r - 7)$

**6.** $g(z) = (z^3 - 1)\left(\dfrac{z^2}{(z^3 - 1)^{1/2}}\right)^2$

**7.** $h(x) = 9(x^6 - 5x^4 + 3x^2 + 12)^3$

**8.** $f(x) = 2\left(\dfrac{1}{5}x^{20} + \dfrac{1}{3}x^{18}\right)^5\left(\dfrac{3}{2}x - \dfrac{5}{2}\right)$

**9.** $y = \dfrac{5t^{1/3} + 2t^{1/2} + 6t^2}{14\sqrt[3]{t}}$

**10.** $y = x^{1/2}\sqrt{x + 4}$

**11.** $f(x) = \left(4x^3 - \dfrac{1}{2}\sqrt{x^3} + 6\right)\left(\dfrac{1}{3}x^6\right)$

**12.** $h = (x^2 + 1)\left(\dfrac{x - 1}{x + 1}\right)$

**13.** $w = \left(\dfrac{4n^3 + 6n^2 + 9}{2\sqrt{n^2}}\right)^5$

**14.** $f(m) = (m^2 + 2m - 9)^{10}(-m + 8)^5$

**15.** $k(p) = \sqrt[3]{p^2 + 6p + 8}\left(\dfrac{p}{p+1}\right)$

**16.** $g(x) = \sqrt{\dfrac{2x+1}{x^2 - 4}}$

**17.** $h(r) = \sqrt[4]{(r^4 + 3)(r^2 + 2r + 9)}$

**18.** $y = \left(\dfrac{x^2 - x + 10}{2x - 1}\right)^3$

**19.** $f(x) = \left(2x - \dfrac{x^3}{4x^2 + 12x - 7}\right)^3$

**20.** $h(t) = \sqrt{1 + \sqrt{t}}$

- Use the Chain Rule in HW6.2.

## HW6.2

**1.** Use the Chain Rule to differentiate the following functions:

    **a.** $(g \circ f)(x)$, where $g(x) = \dfrac{1}{x}$ and $f(x) = (x + 9)^{10}$

    **b.** $(g \circ f)(x)$, where $g(x) = (x + 9)^{10}$ and $f(x) = \dfrac{1}{x}$

**2.** Express each of the following functions as the composition of two functions. Use the Chain Rule to find the derivative of the composition.

    **a.** $g(z) = \left(z^2 - \dfrac{1}{z^2}\right)^6$

    **b.** $p(s) = 1/(8 - 5s + 7s^2)^5$

    **c.** $F(x) = \left(\dfrac{3x^2 - 5}{2x^2 + 7}\right)^2$

    **d.** $m(w) = (\sqrt{w} - 1)^{-2}$

**3.** Suppose $f$ and $g$ are functions such that

$$f(2) = 4, \quad f'(2) = 3, \quad f'(4) = -3;$$
$$g(2) = -1, \quad g'(2) = -2, \quad g'(4) = 5.$$

Find $h'(2)$ when

    **a.** $h(x) = g(f(x))$

    **b.** $h(x) = \dfrac{(f(x))^2}{g(x)}$

    **c.** $h(x) = (f(x))^{-3/2}$

- Use the First Derivative Test to find local extrema in HW6.3.

**HW6.3** The First Derivative Test (see comments following Task 5-5) provides a method for finding the local extrema of a function without knowing the shape of the graph of the function. The general approach involves finding the sign chart for the derivative. Since the value of the derivative gives the slope of the tangent line, the sign chart can be used to determine the locations of the function's horizontal tangents and its local extrema, and the intervals where the function is increasing and decreasing. Based on this information, you can sketch the general shape of the function.

    Use the First Derivative Test to sketch the graphs of the following functions.

    **i.** Find an expression for $f'$ by using the differentiation rules.

    **ii.** Find the sign chart for $f'$ by using the expression for $f'$.

**iii.** Indicate on the sign chart the intervals where $f$ is increasing and decreasing. Indicate the locations of the local extrema of $f$ by applying the First Derivative Test.

**iv.** Find the $(x, y)$-coordinates of the local extrema.

**v.** Plot the local extrema. Sketch the graph of the function using the information on the sign chart for $f'$.

Show all your work.

**1.** $f(x) = x^3 - 12x + 2$

**2.** $f(x) = x^3 + 2$

**3.** $f(x) = x^4 + \dfrac{8}{3}x^3 - 6x^2 + 20$

**4.** $f(x) = \begin{cases} \sqrt{x} & \text{if } x \geq 0 \\ \sqrt{-x} & \text{if } x < 0 \end{cases}$

- Find a general equation for a tangent line in HW6.4.

**HW6.4** Suppose $f$ is a differentiable function defined on an interval containing $x = a$. Find an equation for the tangent line to the graph of $f$ at $P(a, f(a))$.

- Prove two differentiation rules in HW6.5.

**HW6.5**

**1.** Use the Product Rule to prove the Quotient Rule. That is, use the Product Rule to show that

$$\frac{d}{dx}\left( \frac{f(x)}{g(x)} \right) = \frac{g(x)f'(x) - f(x)g'(x)}{(g(x))^2}.$$

Steps:

**i.** Express the quotient $\dfrac{f(x)}{g(x)}$ as the product of two functions. Then use the Product Rule in conjunction with the Extended Power Rule to find the derivative.

**ii.** Show that the result (from part i) equals $\dfrac{g(x)f'(x) - f(x)g'(x)}{(g(x))^2}$ (which is the Quotient Rule).

**2.** Use the limit-based definition of derivative and the Binomial Theorem to show that the Power Rule holds for any positive integer. That is, use the definition of derivative and the Binomial Theorem to prove that

$$\frac{d}{dx}(x^n) = nx^{n-1}, \text{ when } n \text{ is a positive integer.}$$

Steps:

**i.** Use the definition of derivative to express $f'(x)$ as a limit.

**ii.** Let $f(x) = x^n$. Evaluate $f$ at $x$ and $x + h$.

**iii.** According to the Binomial Theorem, if $n$ is a positive integer, then

$$(x + h)^n = x^n + nx^{n-1}h + \frac{n(n - 1)}{2!}x^{n-2}h^2 + \cdots + nxh^{n-1} + h^n.$$

Substitute for $(x + h)^n$, noting that each of the missing terms—indicated by the three dots—contains a power of $h$, namely $h^3$, $h^4$, and so on.

**iv.** Simplify the result, eliminating the $h$ in the denominator.

**v.** Take the limit as $h \to 0$. You should get $nx^{n-1}$.

## SECTION 2

## Analyzing Functional Behavior

The derivative of a function provides you with information about the shape of the graph of the underlying function. By looking at the sign chart for the first derivative, you can determine where the function is increasing and decreasing and you can find the locations of the function's horizontal tangents and local extrema. The first derivative, however, does not give any information about the *concavity* of a function. For instance, although the following functions increase and decrease over the same intervals and they all have the same local maximum, the shapes of their graphs are different since their concavity varies.

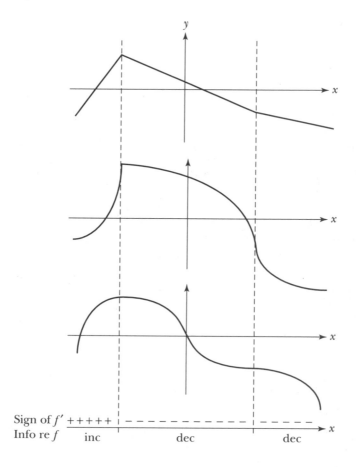

In Unit 1, you discovered that you can determine the concavity of a function by examining the following:

- The location of the graph of the function with respect to the tangent line. Does the graph lie above or below the tangent line?
- The behavior of the slope of the tangent line—or the derivative—as the tangent line travels along the curve from left to right. Is the derivative of the function increasing or decreasing?

In addition, you can determine the concavity of a function by examining

- The sign of the derivative of the derivative, or the second derivative. Is the second derivative positive, negative, or zero?

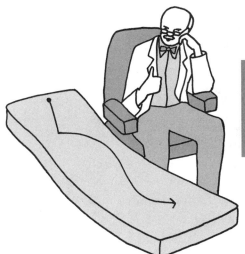

The goals of the next two tasks are to help you recall some observations you made in Unit 1 concerning concavity, to examine how higher-order derivatives provide information about concavity, to explore the connections among the various ways of thinking about concavity, and, finally, to use this information to analyze the behavior of a function—without knowing what its graph looks like ahead of time.

## Task 6-8: Contemplating Concavity (Again)

1. For each of the three graphs given above, label the intervals where the function is concave up and the intervals where it is concave down. Indicate the location of each point of inflection, recalling that an inflection point is where the concavity changes from up to down or vice versa.

2. Examine what happens when a smooth function is concave up. To help see what is going on, it may help if you let your pencil play the role of the tangent line and travel along the curve.

   a. On the axes given below, sketch two smooth curves, one which is concave up and increasing and the other which is concave up and decreasing.

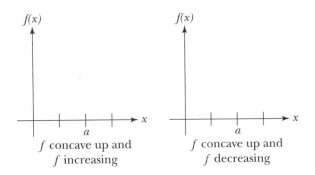

*f* concave up and          *f* concave up and
*f* increasing              *f* decreasing

**b.** On each curve, sketch a few tangent lines.

**c.** Examine the relationship between a function that is concave down and the slope of the tangent line to the curve—that is, examine the relationship between $f$ and $f'$. Summarize your observations by circling the appropriate phrase in each of the following statements.

(1) When $f$ is concave up, the graph of $f$ (lies above, lies below, passes through) the tangent line.

(2) When $f$ is concave up, the slope of the tangent line (is increasing, is decreasing, changes from increasing to decreasing or vice versa) as it travels along the graph from left to right.

(3) When $f$ is concave up, $f'$ (is increasing, is decreasing, has a local extremum).

2. Similarly, examine what happens when a smooth function is concave down.

**a.** On the axes given below, sketch two smooth curves, one which is concave down and increasing and the other which is concave down and decreasing.

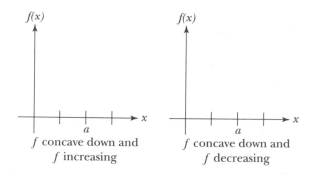

**b.** On each curve, sketch a few tangent lines.

**c.** Examine the relationship between a function that is concave down and the slope of its tangent line—that is, examine the relationship between $f$ and $f'$. Summarize your observations by circling the appropriate phrase in the following statements.

(1) When $f$ is concave down, the graph of $f$ (lies above, lies below, passes through) the tangent line.

(2) When $f$ is concave down, the slope of the tangent line (is increasing, is decreasing, changes from increasing to decreasing or vice versa) as it travels along the graph from left to right.

(3) When $f$ is concave down, $f'$ (is increasing, is decreasing, has a local extremum).

4. Finally, examine what happens at an inflection point.

   **a.** On the axes given below, sketch two smooth curves which have an inflection point at $x = a$, where one is increasing and the other is decreasing.

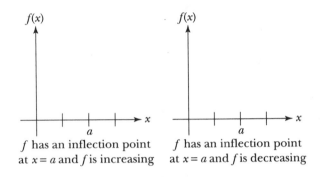

   $f$ has an inflection point   $f$ has an inflection point
   at $x = a$ and $f$ is increasing   at $x = a$ and $f$ is decreasing

   **b.** On each curve, sketch a few tangent lines.

   **c.** Examine the relationship between a function and the slope of the tangent line to the curve—that is, between a function $f$ and its derivative $f'$—at an inflection point. Summarize your observations by circling the appropriate phrase in the following statements.

   (1) When $f$ has an inflection point at $x = a$, the graph of $f$ (lies above, lies below, passes through) the tangent line at $x = a$.

   (2) When $f$ has an inflection point at $x = a$, the slope of the tangent line (is increasing, is decreasing, changes from increasing to decreasing or vice versa) at $x = a$.

   **d.** When $f$ has an inflection point at $x = a$, $f'$ has a local extremum at $x = a$.

   (1) Explain why this is true.

   (2) For each of the following graphs of $f$, determine whether $f'$ has a local maximum or a local minimum at the inflection point $x = a$. Justify your choice based on whether $f'$ is increasing or decreasing to the left or to the right of $x = a$. Sketch a graph of $f'$ near $x = a$.

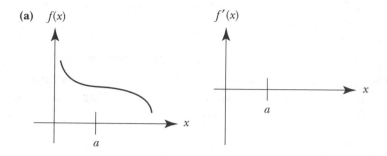

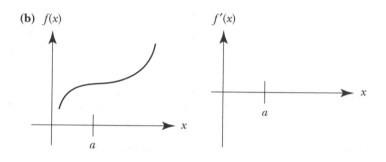

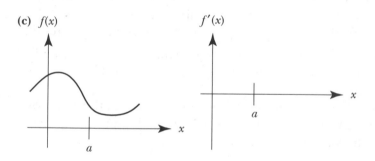

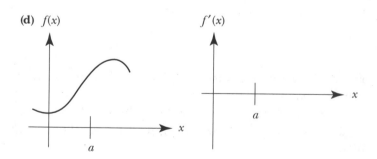

Based on the last task, you can find the intervals where a function is concave up/down by identifying the intervals where the function's derivative is increasing/decreasing, and you can find the location of the function's inflection points by identifying the places where the derivative changes from increasing to decreasing or vice versa.

In general, you can determine when a function is increasing or decreasing by examining its derivative. Since the derivative of a function is a function, you should be able to determine where a derivative is increasing or decreasing by examining its derivative—that is, by examining *the derivative of the derivative*. In the next task, you will calculate some *higher-order derivatives* and begin to think about the information they provide about the shape of the graph of the underlying function.

Before you begin, examine what is meant by "the derivative of the derivative." Consider, for example, the function

$$f(x) = x^3 + 2x + 6.$$

Then, the derivative of $f$ is

$$f'(x) = \frac{d}{dx}(f(x)) = 3x^2 + 2.$$

Since $f'$ is a function, you can find its derivative—which can be denoted by $f''$ to indicate that you have taken the derivative twice:

$$f''(x) = \frac{d}{dx}(f'(x)) = 6x.$$

In general,

- $f'$ is read "$f$ prime" and denotes the first derivative of $f$.
- $f''$ is read "$f$ double prime" and denotes the *second derivative* of $f$ or the derivative of $f'$.
- $f'''$ is read "$f$ triple prime" and denotes the *third derivative* of $f$ or the derivative of $f''$, and so on, where $f^{(n)}$ denotes the $n$th derivative of $f$.

## Task 6-9: Utilizing Higher-Order Derivatives

1. For each of the following functions, find the first, second, and third derivatives.

   *Note: After finding a derivative, simplify its expression before differentiating again.*

   **a.** $f(x) = 3x^5 = 8\sqrt{x} - 4x^{-2}$

**b.** $g(t) = t(8 + t)^4$

**c.** $h(r) = \dfrac{3r - 1}{2r + 3}$

**2.** Consider $p(q) = \dfrac{6}{5}q^5 - 16q^3 + 2q - 100$.

    **a.** Solve $p''(q) = 0$.

    **b.** Evaluate $p'''(-1)$.

    **c.** Find the smallest integer $n$ such that $p^{(n)}(q) = 0$ for any real number $q$.

**3.** Investigate the connection between second derivative of a function, the behavior of the first derivative of the function, and the concavity of the function.

Consider $f(x) = \frac{1}{12}x^4 - 2x^2 + x - 2$. Find the sign chart for $f''$ as you complete parts a through c given below.

Sign of $f''$

$\longrightarrow x$

Info re $f'$

Info re $f$

**a.** Find the Sign of $f''$ by applying the same approach you used to investigate the sign of the first derivative.

    **(1)** Find an expression for $f''(x)$.

    **(2)** Find the $x$-values where $f''(x) = 0$.

    **(3)** Identify the intervals where $f''$ is positive and negative.

    **(4)** Record the information on the sign chart for $f''$ by placing 0's, +'s, and −'s above the axis to indicate where $f''(x) = 0$, $f''(x) > 0$, and $f''(x) < 0$.

**b.** Fill in the Info re $f'$ line on the sign chart for $f''$, labeling the intervals where $f'$ is increasing with "inc" and those where $f'$ is decreasing with "dec." Mark the $x$-values of the local extrema of $f'$ with "LM" for local maximum and "lm" for local minimum.

*Note: A function is increasing when its derivative is positive, and decreasing when its derivative is negative; the function has a local extrema at a point where it changes from increasing to decreasing. In particular, f' is increasing when its derivative f" is positive, and f' is decreasing when f" is negative.*

**c.** Fill in the Info re $f$ line on the sign chart for $f''$ labeling the intervals where $f$ is concave up with "CU" and those where $f$ is concave down with "CD." Mark the $x$-values of the inflection points of $f$ with "IP."

*Note: Based on the observations you made in Task 6-8, a function f is concave up when its first derivative f' is increasing, and f is concave down when f' is decreasing.*

**d.** Find the $y$-coordinate of each inflection point by evaluating $f$ at the corresponding $x$-value.

You are ready to pull this all together, but first let's summarize our observations. Suppose $f$ is a continuous function. Then, the sign chart for the first derivative $f'$ indicates the intervals where $f$ is increasing and decreasing and gives the $x$-coordinates of the local extrema of $f$. In particular,

- $f$ is increasing when $f'(x) > 0$.
- $f$ is decreasing when $f'(x) < 0$.
- $f$ has a local extremum when $f'$ changes sign—that is, when $f$ changes from increasing to decreasing or vice versa. If $f$ has a local extremum at $x = c$, then $f'(c) = 0$ or $f'(c)$ does not exist.

The sign chart for the second derivative $f''$ indicates the intervals where the derivative $f'$ is increasing and decreasing and the location of the local extrema of $f'$. It also indicates the intervals where $f$ is concave up and concave down and gives the $x$-coordinates of the inflection points of $f$. In particular,

- $f'$ is increasing when $f''(x) > 0$.
- $f'$ is decreasing when $f''(x) < 0$.
- $f'$ has a local extremum when $f''$ changes sign—that is, when $f'$ changes from increasing to decreasing or vice versa. If $f'$ has a local extremum at $x = c$, then $f''(c) = 0$ or $f''(c)$ does not exist.
- $f$ is concave up when $f''(x) > 0$.
- $f$ is concave down when $f''(x) < 0$.
- $f$ has an inflection point when $f''$ changes sign—that is, when $f$ changes from concave up to concave down or vice versa. If $f$ has an inflection point at $x = c$, then $f''(c) = 0$ or $f''(c)$ does not exist.

## Task 6-10: Sketching Curves

1. For each of the following functions $f$:

   i. Find the sign chart for the first derivative $f'$.

   - Determine the intervals where $f$ is increasing and decreasing.
   - Find the $x$-coordinate of each turning point of $f$ and determine if it is a local maximum or minimum.
   - Find the $y$-coordinate of each local extremum of $f$.

   ii. Find the sign chart for the second derivative $f''$.

   - Determine the intervals where $f'$ is increasing and decreasing.
   - Determine the intervals where $f$ is concave up and concave down.
   - Find the $(x, y)$-coordinates of the inflection points of $f$.

   iii. Use the information from parts i and ii to sketch the graph of $f$.

**a.**   $f(x) = -x^3 + 3x - 2$

**b.**   $h(t) = 3t^4 - 4t^3 + 6$

2. For each of the following sets of conditions, sketch the graph of a function satisfying all the conditions. Before sketching a graph of the function, determine what information the sign charts give about the shape of the graph.

**a.** $f$ is continuous for all real numbers.

$f(-5) = 0$

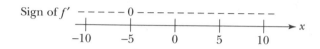

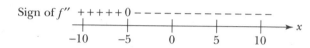

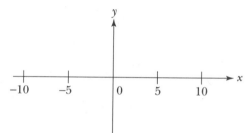

**b.** $h$ is continuous for all real numbers.

$h'(-3) = 0$ and $h'(1) = 0$

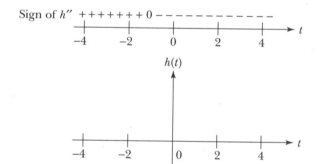

**c.** $r$ is continuous for all real numbers.

$r''(p) = p^3$

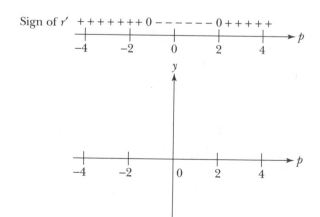

**d.** $f$ is continuous for all real numbers.

$f'(x) = 0$ for $x = 0$, 4, and 8

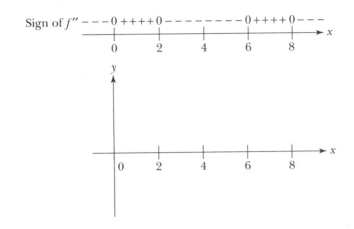

In addition to local extrema, a function may have *absolute extrema*. In particular, if $f$ is a function defined on an interval $I$, then

- $f$ has an *absolute minimum* $f(a)$ at $x = a$ if $f(a) \le f(x)$, for all $x$ in $I$.
- $f$ has an *absolute maximum* $f(a)$ at $x = a$ if $f(a) \ge f(x)$, for all $x$ in $I$.

A function may have an absolute maximum or minimum at more than one $x$ value in $I$, or it may have an absolute maximum but not an absolute minimum and vice versa. The interval $I$ does not have to include its endpoints. Think about absolute extrema as you do the next task.

## Task 6-11: Locating Absolute Extrema

1. Absolute extrema may occur where the function has a turning point or at an endpoint of the interval $I$. Suppose $I$ is the closed interval $[-2, 4]$. For each of the following lists of properties, sketch a graph of a continuous function over $I$ having the given properties.

   **a.** $f$ has three turning points. $f$ has an absolute minimum at the left endpoint of $I$. $f$ has an absolute maximum at the rightmost turning point.

   **b.** $f$ has no turning points. $f$ has an absolute maximum at the left endpoint of $I$.

    **c.** The absolute extrema of $f$ do not occur at the endpoints of $I$.

2. Let $f$ be a differentiable function defined on an interval $I$. In parts a through d:

    **i.** Determine the location of the absolute minima of $f$ (if one exists).

    **ii.** Determine the location of the absolute maxima of $f$ (if one exists).

    **a.** Suppose $f$ is strictly increasing on $I$, and

        **(1)** $I = [b, c]$, where $b$ and $c$ are real numbers.

        **(2)** $I = [b, \infty)$, where $b$ is a real number.

    **b.** Suppose $f$ is strictly decreasing on $I$, and

        **(1)** $I = [b, \infty)$, where $b$ is a real number.

        **(2)** $I = (-\infty, c]$, where $c$ is a real number.

    **c.** Suppose $f$ is concave up on $I$, $f'(x) = 0$ for exactly one value of $x$ in $I$ which is not an endpoint of $I$, and

        **(1)** $I = (-\infty, c]$, where $c$ is a real number.

        **(2)** $I = (-\infty, \infty)$.

    **d.** Suppose $f$ is concave down on $I$, $f'(x) = 0$ for exactly one value of $x$ in $I$ which is not an endpoint of $I$, and

        **(1)** $I = [b, c]$, where $b$ and $c$ are real numbers.

        **(2)** $I = (-\infty, \infty)$.

**3.** Suppose $f$ is a continuous function defined on a closed interval $I$. Suppose you know the $x$-coordinates of all the possible turning points of $f$ that lie within the interval $I$. In other words, you know all the values of $x$ in $I$, such that either $f'(x) = 0$ or $f'(x)$ does not exist. Describe how you can use this information to find the absolute extrema of $f$, without considering the sign charts for either $f'$ or $f''$ and without sketching the graph of $f$.

**4.** Hopefully you observed in part 3, that a quick way to locate the absolute extrema of $f$ in a closed interval $I$ is to:

- Find the $x$-coordinates of all possible turning points of $f$ that lie within the interval $I$.
- Evaluate $f$ at each of the possible turning points and at the endpoints of $I$.
- Compare the resulting values of $f(x)$. The largest value gives the location(s) of the absolute maxima of $f$, and the smallest value gives the location of the absolute minima of $f$.

Use this approach to find the absolute extrema of the function $f(x) = -4x^2 - 4x + 5$, on the following intervals:

    **a.** $[-2, 0]$

    **b.** $[-3, -1]$

**5.** Given a rectangular piece of cardboard, 16" wide and 21" long, what size square would you cut from each corner to construct a box with maximum volume?

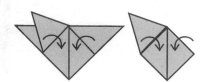
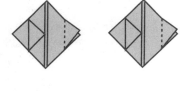
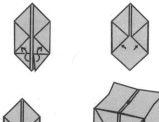

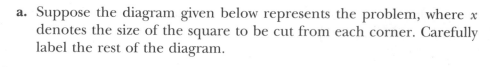

**a.** Suppose the diagram given below represents the problem, where $x$ denotes the size of the square to be cut from each corner. Carefully label the rest of the diagram.

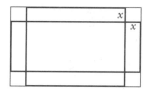

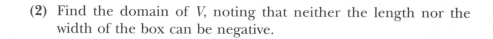

**b.** Suppose $V$ is the volume of the box.

   **(1)** Find an equation for $V$ in terms of $x$.

   **(2)** Find the domain of $V$, noting that neither the length nor the width of the box can be negative.

**c.** The goal of this problem is to use calculus to find the size of the cutout $x$ that maximizes the volume $V$—that is, to find a value of $x$ where $V(x)$ is an absolute maximum.

   **(1)** Find the $x$-coordinates of all possible turning points of $V$ that lie in the domain of $V$.

   **(2)** Evaluate $V$ at the possible turning points and at the endpoints of its domain. Find the value of $x$ where $V$ has an absolute maximum by comparing the results.

**d.** Give the dimensions of the box that maximize the volume.

## Unit 6 Homework After Section 2

• Complete the tasks in Section 2. Be prepared to discuss them in class.

• Summarize what the signs and zeros of the first and second derivatives says about the behavior of a function in HW6.6.

**HW6.6** Fill in the following table. To help you get started, the first row is completed.

| Signs of Derivatives | Effect on the Graph of $f$ at $x = a$ | Sketch of the Graph of $f$ near $x = a$ |
|---|---|---|
| **1.** $f'(a) > 0$ <br> $f''(a) > 0$ | $f$ is increasing at $x = a$. <br> $f'$ is increasing at $x = a$. <br> $f$ is concave up at $x = a$. | |
| **2.** $f'(a) > 0$ <br> $f''(a) < 0$ | | |
| **3.** $f'(a) < 0$ <br> $f''(a) > 0$ | | |
| **4.** $f'(a) < 0$ <br> $f''(a) < 0$ | | |
| **5.** $f'(a) = 0$ <br> $f''(a) > 0$ | | |
| **6.** $f'(a) = 0$ <br> $f''(a) < 0$ | | |

• Scrutinize the information given by the first and second derivatives of a function in HW6.7.

**HW6.7**

1. In parts a–d, sketch a graph that satisfies the listed criteria.

   **a.** $f$ is a continuous function defined for all real numbers. $f(0) = 0$. The sign graphs of the first and second derivatives of $f$ are

**b.** $k$ is a continuous function defined for all real numbers. $k(0) = 0$. The sign graphs of the first and second derivatives of $k$ are

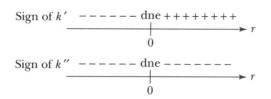

**c.** $f$ is a continuous function defined for all real numbers. $f$ has local extrema at $x = 0, 2, 4, 6,$ and 8. $f'(0) = f'(4) = f'(8) = 0$. $f'(2)$ and $f'(6)$ do not exist. $f$ is always concave up.

**d.** $s$ is a continuous function defined for all real numbers. $s(x) < 0$, $s'(x) < 0$, and $s''(x) < 0$ for all $x$.

**2.** Analyze the following graphs.

   **a.** Consider the following graph of $f'$.

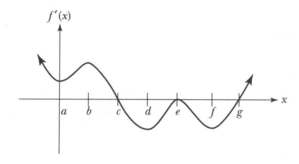

   **(1)** Use the labeled points on the graph of $f'$ to identify the values of $x$ where the graph of $f$ has

   **(a)** a local maximum

   **(b)** a local minimum

   **(c)** a horizontal tangent that is not a local extremum

   **(d)** an inflection point

   **(2)** Use the labeled points to identify the intervals where the graph of $f$ is

   **(a)** concave up

   **(b)** concave down

**b.** Consider the following graph of $f''$.

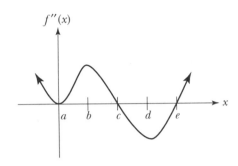

(1) Use the labeled points on the graph of $f''$ to identify the values of $x$ where the graph of $f$ has an inflection point.

(2) Use the labeled points to identify the intervals where the graph of $f$ is

(a) concave up

(b) concave down

**c.** Consider the following graph of $f$.

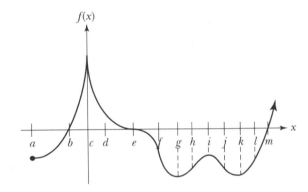

(1) Use the labeled points on the graph of $f$ to identify the intervals where

(a) $f(x) > 0$

(b) $f'(x) > 0$

(c) $f''(x) > 0$

(2) Use the labeled points to identify the values of $x$ where

(a) $f'(x)$ does not exist

(b) the graph of $f''$ crosses the $x$-axis

(c) $f$ has an absolute extremum

(d) both $f'(x)$ and $f''(x)$ equal zero

**3.** Suppose $f'$ is continuous and $f'(x) = 0$ for exactly two values of $x$. Explain why it is not possible for $f$ to have the following values: $f(-2) = f(0) = f(2) = 2$ and $f(-1) = f(1) = -1$.

4. The graph of $f(x) = ax^2 + bx + c$ is a parabola, for all $a$, $b$, and $c$ real numbers and $a \neq 0$. Use calculus to show that the vertex of a parabola is at $x = -\left(\dfrac{b}{2a}\right)$.

• Suppose a function models a situation. Determine what the shape of the function tells you about the situation in HW6.8.

**HW6.8** Recall that the first derivative gives the rate of change of the function. Consequently, if the function is increasing, its rate of change is positive, and if the function is decreasing, its rate of change is negative. Moreover, if the function is concave up, its rate of change is increasing, and if the function is concave down, its rate of change is decreasing.

Assume each of the following graphs model the profits of a company over time. Describe what the graphs tell you about the change in the company's profits over time. When are the profits growing? Decaying? Holding steady? When is the rate of growth/decay increasing? Decreasing? Holding steady?

For example, in part 1, since $f$ is increasing, your rate of change is positive and your profits are increasing as time goes by. Moreover, since $f$ is concave up, the rate of change of your profits is increasing. In other words, as time passes, your profits are growing and they are growing at a faster and faster rate. This sounds like a good position to be in!

1.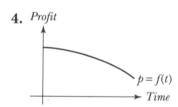

2.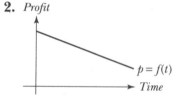

3.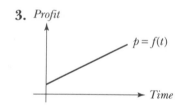

4.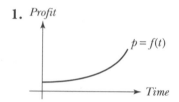

5.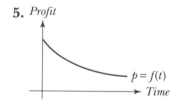

6.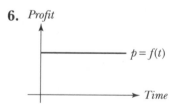

7.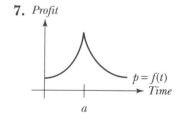

8.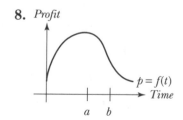

**9.** *Profit*

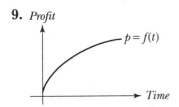

**10.** *Profit*

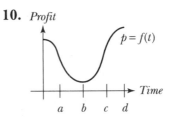

• Analyze the behavior of some functions in HW6.9.

**HW6.9** For each of the following functions:

   **i.** Find the sign chart for the first derivative.

     • Determine the intervals where the function is increasing and decreasing.
     • Find the location of each turning point and determine if it is a local maximum or minimum.
     • Find the $(x, y)$-coordinates of the local extrema of the function.

   **ii.** Find the sign chart for the second derivative.

     • Determine the intervals where the function is concave up and concave down.
     • Find the $(x, y)$-coordinates of the inflection points of the function.

   **iii.** Sketch a graph of the function

**1.** $w(x) = x^2 + x - 1$         **4.** $m(t) = t^4 - 2t^3$

**2.** $f(x) = (x - 1)^3$         **5.** $f(r) = r^4 - 2r^2$

**3.** $s(y) = 3 - y^4$          **6.** $t(x) = 8x^3 + 48x^2 + 96x + 64$

• Consider some alternate notations for higher order derivatives in HW6.10.

**HW6.10** The general derivative of a function is itself a function, which typically is represented by an expression. The following chart summarizes the different notations for the general derivative used in this section.

| Function | 1st Derivative | 2nd Derivative | nth Derivative |
|---|---|---|---|
| $y$ | $y'$ <br> $\dfrac{dy}{dx}$ | $y''$ <br> $\dfrac{d^2y}{dx^2}$ | $y^{(n)}$ <br> $\dfrac{d^ny}{dx^n}$ |
| $f(x)$ | $f'(x)$ <br> $\dfrac{d}{dx}(f(x))$ | $f''(x)$ <br> $\dfrac{d^2}{dx^2}(f(x))$ | $f^{(n)}(x)$ <br> $\dfrac{d^n}{dx^n}(f(x))$ |

    If you evaluate the general derivative at a particular value, you get a number, which tells you the rate of change of the function—or the slope of the tangent line—at that value. The table below shows the corresponding ways to indicate the value of a derivative at a particular input value, say $x = a$.

| Function Value at $x = a$ | 1st Derivative Value at $x = a$ | 2nd Derivative Value at $x = a$ | nth Derivative Value at $x = a$ |
|---|---|---|---|
| $y\|_{x=a}$ | $y'\|_{x=a}$ $\left.\dfrac{dy}{dx}\right\|_{x=a}$ | $y''\|_{x=a}$ $\left.\dfrac{d^2y}{dx^2}\right\|_{x=a}$ | $y^{(n)}\|_{x=a}$ $\left.\dfrac{d^ny}{dx^n}\right\|_{x=a}$ |
| $f(a)$ | $f'(a)$ $\left.\dfrac{d}{dx}(f(x))\right\|_{x=a}$ | $f''(a)$ $\left.\dfrac{d^2}{dx^2}(f(x))\right\|_{x=a}$ | $f^{(n)}(a)$ $\left.\dfrac{d^n}{dx^n}(f(x))\right\|_{x=a}$ |

To find the value of a derivative at a specified input, first find the general derivative and then evaluate the result at the given value. For example, to evaluate $\left.\dfrac{dy}{dx}\right|_{x=6}$ when $y = x^3$, first find the derivative: $\dfrac{dy}{dx} = \dfrac{d}{dx}(x^3) = 3x^2$. Next, evaluate the derivative at $x = 6$: $\left.\dfrac{dy}{dx}\right|_{x=6} = 3 \cdot 6^2 = 108$.

1. Consider $y = (x^2 - 4)^5$.

   a. Find $y'\|_{x=1}$.

   b. Find $\left.\dfrac{d^2y}{dx^2}\right|_{x=0}$.

   c. Find all the values of $x$ such that $y' = 0$.

2. Consider $g(t) = \sqrt{t} + t^2 + 101$.

   a. Find $\left.\dfrac{d^2}{dt^2}(g(t))\right|_{t=16}$.

   b. Find $g^{(4)}(1)$.

   c. Find the domain of $g'$.

- If a function is sufficiently "nice," the second derivative may also provide a quick way to locate the local extrema of a function. Develop a statement of the Second Derivative Test and use it to identify local extrema in HW6.11.

**HW6.11 The Second Derivative Test.**

1. Let $f$ be twice differentiable on an open interval containing $x = c$.

   a. Suppose $f'(c) = 0$ and $f''(c) > 0$.

      (1) Sketch a graph of $f$ for $x$ near $c$.

      (2) Describe the shape of the graph of $f$ at $x = c$.

   b. Suppose $f'(c) = 0$ and $f''(c) < 0$.

      (1) Sketch a graph of $f$ for $x$ near $c$.

      (2) Describe the shape of the graph of $f$ at $x = c$.

**2.** Based on your observations in part 1, to find the candidates for the local extrema of a function that is twice differentiable, locate all the horizontal tangents and then find the sign of the second derivative at each of these values. If the second derivative is positive, the function is concave up and, consequently, it has a local minimum at the horizontal tangent. If the second derivative is negative, the function is concave down and, consequently, it has a local maximum at the horizontal tangent. In other words:

Suppose $f$ is twice differentiable on an open interval containing $c$ where $f'(c) = 0$. Then

- $f$ has a local minimum at $x = c$, if $f''(c) > 0$,
- $f$ has a local maximum at $x = c$, if $f''(c) < 0$.

This is called the *Second Derivative Test*.

Note that the test gives no information about how $f$ behaves when $f''(c) = 0$; the function may have a local maximum, a local minimum, or neither. In this case, you have to use the First Derivative Test.

Use the Second Derivative Test to identify the local extrema of the functions given in parts a and b below. For each function, find the $x$-values where $f'(x) = 0$ and then find the sign of $f''$ at each of these values.

**a.** $f(x) = \frac{8}{3}x^3 - 2x + \frac{1}{3}$

**b.** $h(t) = 6 + t^2 - \frac{1}{2}t^4$

**3.** The Second Derivative Test does not always apply.

**a.** Consider $f(x) = (x - 1)^4 + 2$.

  **(1)** Show why the Second Derivative Test does not apply.

  **(2)** Use the First Derivative Test to determine the shape of the graph of $f$ at the horizontal tangent.

**b.** Sometimes you can use the Second Derivative Test to determine the behavior of a function at some horizontal tangents, but not at others. Consider for instance, $f(x) = x^5 - 15x^3$.

  **(1)** Show that the Second Derivative Test can be used to classify two of the places where $f$ has a horizontal tangent, but not the third.

  **(2)** Use the First Derivative Test to determine the shape of the graph of $f$ at the remaining horizontal tangent.

- Solve some optimization problems in HW6.12 to HW6.14.

**HW6.12** You have decided to go into business for yourself making hand-held computers that do all the calculations, symbolic manipulations, and graphing needed for Workshop Calculus (no more of this pencil and paper stuff!). To help determine the market potential of your product, you hire a marketing firm to survey students at your school. They conclude that in order to sell $x$ hand-held computers, the price per minicomputer must be

$$D(x) = 280 - 0.4x.$$

They also conclude that the total cost of producing $x$ minicomputers is given by

$$C(x) = 8000 + 0.6x^2.$$

1. Find the price you must charge for each unit in order to sell 50 minicomputers. Find the price for 120. Find the price for 200.

2. Find your total costs for producing 50, 120, and 200 minicomputers.

3. Your goal, obviously, is to maximize your profits. In order to calculate your profits, however, you need to know how much revenue you will be realizing. Find an expression for the total revenue $R(x)$, noting that

   total revenue = (number of units) · (price per unit).

   Use your expression for $R(x)$ to calculate your total revenue if you sell 50 minicomputers; 120 minicomputers; 200 minicomputers.

4. Find an expression for your total profit $P(x)$, noting that

   total profit = total revenue − total cost.

   Use your expression for $P(x)$ to calculate your total profit if you sell 50 minicomputers; 120 minicomputers; 200 minicomputers.

5. Determine how many minicomputers you must sell in order to maximize your profits.

   *Hint: Consider the sign graph for P′.*

6. Find your maximum profit.

7. Find the price per minicomputer that you must charge in order to maximize your profits.

**HW6.13** As one who is deeply interested in the environment, you have decided that cars are out and public transportation is in. The only snag is that the Greater Carlisle Area does not have its own transportation system. "No problem," you say. "I'll start my own company." After surveying the situation, you conclude that if you charge a fare of 60¢, you will average 5000 riders per day, but for every fare increase of 10¢, you will lose 500 customers. Although your motive for establishing the company is purely altruistic (you want to protect the environment), the business side of you wants to ensure that you maximize the amount of money you realize. So the question is: What fare should you charge in order to maximize your revenue?

Suppose $x$ is the number of 10¢ fare increases (over 60¢). For example, if $x = 3$, then the fare would be 60¢ + 3 · 10¢ or 90¢. If $x$ is negative, the fare decreases.

1. Find an expression for your revenue $R(x)$, where $x$ is the number of 10¢ fare increases.

   *Hint: Find an expression for the number of riders and for the fare (in terms of x) and use the fact that revenue = number of riders · fare.*

**2.** Determine the number of 10¢ fare increases/decreases you need to implement in order to maximize your revenue.

*Hint: Consider the sign graph for R'.*

**3.** Find the fare you need to charge in order to maximize your revenue.

**4.** Estimate the number of riders, assuming you charge the fare that maximizes your revenue.

**HW6.14** As the person in charge of Dickinson College's Big-Little program, you want to fence in a playground, adjacent to the Kline Center, for the smaller participants in the program. You talk the College into giving you 120 yards of fencing. Now the question is: What dimensions will maximize the area of the playground?

*Note: You do not need fencing next to the Kline Center.*

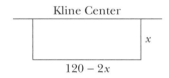

Kline Center

$x$

$120 - 2x$

**1.** Find an expression for the area $A(x)$.

**2.** Find the dimensions that will maximize the area.

**3.** Find the maximum area.

## SECTION 3

## Differentiating Trigonometric, Exponential, and Logarithmic Functions

In Section 1, you developed rules for differentiating combinations of functions, including

- the scalar multiple of a function
- the sum and difference of two functions
- the product of two functions
- the quotient of two functions
- the composition of two functions

You used these rules to find the derivatives of polynomial, rational, and power functions. In this section, you will extend this list of basic functions to include trigonometric, exponential, and logarithmic functions.

The goal of the first task is to discover the derivatives of the sine and cosine functions. In each case, you will identify the derivative—which happens to be familiar function—by examining the behavior of the slope of the tangent line as it travels along the graph. Then, you will use the Chain Rule to find the derivative of more complex trigonometric functions, such as

$$f(\theta) = \sin(2\theta + \pi) \quad \text{or} \quad g(t) = \cos\left(\frac{1}{t}\right).$$

## Task 6-12: Finding the Derivatives of sin(x) and cos(x)

1. Find the derivative of $f(x) = \sin(x)$.

   **a.** Consider the following graph of $f(x) = \sin(x)$, $-\frac{\pi}{2} \le x \le \frac{5\pi}{2}$.

   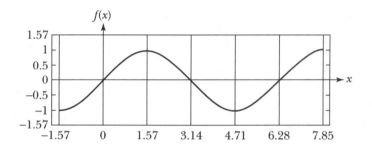

   The horizontal axis of the graph of $f(x) = \sin(x)$ is labeled using decimal values, but it will be easier to work using radians. Relabel the horizontal axis in terms of radians, recalling that $\pi \approx 3.14$.

   **b.** Analyze the shape of the graph of $f(x) = \sin(x)$. Use this information to sketch the graph of $f'$.

   **(1)** Find the sign chart for $f'$ by examining the shape of the graph of $f$. Use the sign chart to determine the following information about $f'$ (Info re $f'$):

   **(a)** The places where the graph of $f'$ intersects the $x$-axis

   **(b)** The intervals where the graph of $f'$ lies above the $x$-axis and the intervals where $f'$ lies below the axis

   **(2)** Find the sign chart for $f''$ by examining the shape of the graph of $f$. Use the sign chart to determine the following information about $f'$ (Info re $f'$):

   **(a)** The intervals where the graph of $f'$ is increasing and where it is decreasing

**(b)**  The locations of the local extrema of $f'$

Sign of $f''$

Info re $f'$    $-\frac{\pi}{2}$      $0$      $\frac{\pi}{2}$      $\pi$      $\frac{3\pi}{2}$      $2\pi$      $\frac{5\pi}{2}$      $x$

**(3)**  At each of the inflection points on the graph of $f$, use a straight-edge to draw the tangent line and then calculate the slope of the line. Based on these calculations, approximate the value of $f'(x)$ at each of the turning points of $f'$.

**(4)**  Use the information from parts (1) to (3) to sketch the graph of the derivative of $f(x) = \sin(x)$.

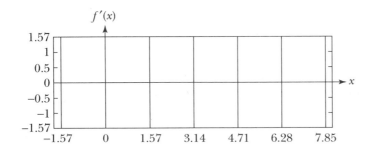

**(5)**  The graph of $f'$ is the graph of a familiar function. Represent the derivative of the sine function by an expression.

**2.**  Use a similar approach to find the derivative of $f(x) = \cos(x)$.

   **a.**  Consider the following graph of $f(x) = \cos(x)$, $-\frac{\pi}{2} \le x \le \frac{5\pi}{2}$. Relabel the horizontal axis in terms of radians.

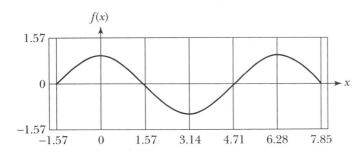

   **b.**  Analyze the shape of the graph of $f(x) = \cos(x)$. Use this information to sketch the graph of $f'$.

**(1)** Find the sign chart for $f'$, and determine the places where the graph of $f'$ intersects the $x$-axis and the intervals where the graph of $f'$ lies above or below the $x$-axis.

**(2)** Find the sign chart for $f''$, and determine the intervals where the graph of $f'$ is increasing and decreasing and the locations of the local extrema of $f'$.

**(3)** Estimate the value of $f'(x)$ at each of the turning points of $f'$ by calculating the slope of the tangent line at each of the inflection points on the graph of $f$.

**(4)** Use the information from parts (1) to (3) to sketch the graph of the derivative of $f(x) = \cos(x)$.

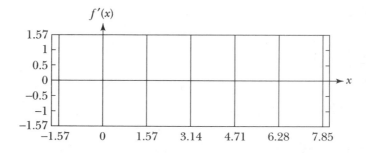

**(5)** The graph of $f'$ is the reflection of a familiar function through the horizontal axis. Represent the derivative of the cosine function by an expression.

In the last task, you discovered that $\dfrac{d}{dx}(\sin(x)) = \cos(x)$ and $\dfrac{d}{dx}(\cos(x)) = -\sin(x)$. In the next task, you will use the Chain Rule to find rules for differentiating $\sin(f(x))$ and $\cos(f(x))$, where $f$ is a differentiable function. You will practice using these new rules in conjunction with the rules you already know to differentiate functions containing trigonometric expressions, such as

$$h(x) = \cos^{10}(x)\,\sin(x^6) \quad \text{or} \quad p(t) = \frac{(t+1)^3}{\sin(t+1)}.$$

For example, to find the derivatives of $h$ and $p$, you would use the Product Rule and the Quotient Rule, respectively, along with the new trigonometric rules.

Before you begin the task, it might be helpful to review the rules for differentiation at the end of Section 1.

## Task 6-13: Finding Derivatives of Functions Containing Trigonometric Expressions

1. Recall that the Chain Rule $\dfrac{d}{dx}((g(f(x)))) = g'(f(x))f'(x)$ is used to differentiate the composition of two differentiable functions $g$ and $f$. Use the Chain Rule to find rules for differentiating $\sin(f(x))$ and $\cos(f(x))$.

   **a.** Consider $\sin(f(x))$, where $f$ is differentiable.

   **(1)** $\sin(f(x))$ can be expressed as the composition of two functions, $g$ and $f$. What is $g(x)$ in this case?

   **(2)** Use the Chain Rule to show that

   $$\frac{d}{dx}(\sin(f(x))) = \cos(f(x)) \cdot f'(x).$$

   **(3)** According to part (2), the derivative of the sine of a function equals the cosine of the function times the derivative of the function. Use this new rule to show that

   $$\frac{d}{dx}(5\sin(x^3 - \pi)) = 15x^2\cos(x^3 - \pi).$$

   **b.** Similarly, consider $\cos(f(x))$, where $f$ is differentiable.

   **(1)** $\cos(f(x))$ can be expressed as the composition of two functions, $g$ and $f$. What is $g(x)$ in this case?

(2) Use the Chain Rule to show that

$$\frac{d}{dx}(\cos(f(x))) = -\sin(f(x)) \cdot f'(x).$$

(3) Thus, the derivative of the cosine of a function equals minus the sine of the function times the derivative of the function. Use the new rule to show that

$$\frac{d}{dx}\left(\cos\left(\frac{1}{x}\right)\right) = x^{-2}\sin\left(\frac{1}{x}\right).$$

2. Find the derivatives of the functions listed below in parts a through f by using the rules for finding derivatives of combinations of functions and the new rules.

$$\frac{d}{dx}(\sin(x)) = \cos(x),$$

$$\frac{d}{dx}(\sin(f(x))) = \cos(f(x)) \cdot f'(x),$$

$$\frac{d}{dx}(\cos(x)) = -\sin(x),$$

$$\frac{d}{dx}(\cos(f(x))) = -\sin(f(x)) \cdot f'(x)$$

Show all your work and state the differentiation rule you are using.

a. $f(x) = \sin(2x^3 - 4x)$

b. $y = \cos\left(\dfrac{x-1}{x+1}\right)$

**c.** $g(r) = \sin(\cos(r))$

**d.** $h(\theta) = \sin(\theta^2)\cos(\theta + \pi)$

**e.** $y = \dfrac{\sin(z^6)}{z^6 - \cos(z)}$

**f.** $y = \sin^2(t^2).$
   Recall that $\sin^n f(x) = (\sin f(x))^n.$

In the homework exercises at the end of this section, you will use the rules for finding the derivatives of sine and cosine to find rules for differentiating the other basic trigonometric functions. In the meantime, think about how to find the derivative of exponential functions and logarithmic functions.

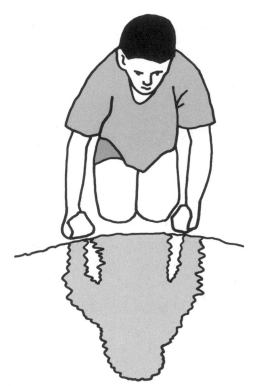

In Unit 3, you examined exponential functions with base $a$.

$$f(x) = a^x \quad \text{for any real number } x,$$

where $a$ is a positive real number. You observed that as the value of the base increases, the corresponding exponential functions grow faster and faster; for example, $3^x$ grows more rapidly than $2^x$.

You also looked at the reflection of an exponential function through the diagonal line $y = x$. However, when you tried to find an expression for the reflection of an exponential function through the line $y = x$ by interchanging $x$ and $y$ in the usual way, you ran into a problem. In particular, interchanging $x$ and $y$ in $y = a^x$ led to $x = a^y$, which was fine except you could not solve this new equation for $y$ algebraically. Consequently, we introduced log notation, where

$$y = \log_a x \quad \text{is equivalent to} \quad x = a^y.$$

We referred to these new functions as *logarithmic functions base a*, where $x = a^y$ is said to be the *exponential form* of $y = \log_a x$.

The domain of $a^x$ is the set of all real numbers and the domain of $\log_a x$ is $x > 0$. You can invert back and forth between the graphs of the associated exponential and logarithmic functions by reflecting the graphs through the line $y = x$.

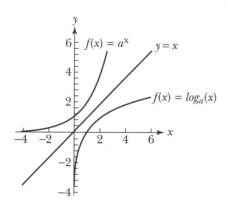

Moreover, since exponential and logarithmic functions are inverses of one another,

$$\log_a(a^x) = x \quad \text{for all } x$$

and

$$a^{\log_a x} = x \quad \text{for all } x > 0.$$

The base $a$ can be any real positive number, but a special value for the base is $e$, where to 15 decimal places

$$e \approx 2.718281828459045.$$

The associated logarithmic function $\log_e(x)$ is denoted by $\ln(x)$ and called the *natural log*. One unanswered question is: What is so special about the number $e$ and the function $f(x) = e^x$? In the next task, you will find the derivative of $e^x$, and in the process discover one reason why $e$ is so unusual.

---

## Task 6-14: Finding the Derivatives of $e^x$ and $e^{f(x)}$

1. What does the graph of $f(x) = e^x$ tell you about the shape of the graph of its derivative? Consider the following graph of $f(x) = e^x$, where $-2 \le x \le 2$:

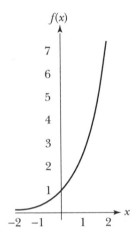

Answer the following questions about the shape of the graph of the derivative of $f(x) = e^x$, using the approach you used to determine the shapes of the graphs of the derivatives of the sine and cosine functions in Task 6-12.

**a.** Find all the $x$-values where the graph of $f'$ intersects the horizontal axis. If it does not intersect the axis, explain why.

**b.** Describe where the graph of $f'$ is located with respect to the horizontal axis (above, below, both above and below). Explain how you know that this is the case.

**c.** Find the intervals where $f'$ is increasing and the intervals where $f'$ is decreasing. Justify your response.

**d.** Find the turning points of $f'$. If $f'$ has no turning points, explain how you know this.

e. Analyze the behavior of the slope of the tangent line as it travels along the graph of $f(x) = e^x$.

(1) Suppose the tangent line travels along the graph from left to right. What happens to the value of the slope of the tangent line as $x$ gets larger and larger? In other words, describe what is happening to the value of $f'(x)$ as $x \to +\infty$.

(2) Similarly, suppose the tangent line travels from right to left. What happens to the value of the slope of the tangent line as $x$ gets more and more negative? In other words, describe what is happening to the value of $f'(x)$ as $x \to -\infty$.

f. Based on you answers to parts a through e, summarize what you know about the location of the graph with respect to the horizontal axis and about the shape of the graph of $f'$.

2. By examining the graph of $f(x) = e^x$ you know that $f'$ is always increasing and you know that the graph of $f'$ is located above the horizontal axis. Moreover, it appears that $f'(x)$ gets larger and larger as $x \to +\infty$ and the values of $f'(x)$ get closer and closer to 0 as $x \to -\infty$. Unfortunately, a lot of functions exhibit this general behavior; so unlike the sine function, there isn't an obvious expression for $f'(x)$.

   Another approach is to use the definition of derivative to find $f'(x)$. Try this.

a. The following sequence of steps shows that

$$f'(x) = e^x \lim_{h \to 0} \frac{e^h - 1}{h}.$$

This will get you part way through the process of representing $f'(x)$ by an expression and finding a rule for differentiating $f(x) = e^x$. Provide an explanation for each step.

| Step | Explanation |
|---|---|

$$f'(x) = \lim_{h \to 0} \frac{f(x + h) - f(x)}{h}$$

$$= \lim_{h \to 0} \frac{e^{(x+h)} - e^x}{h}$$

$$= \lim_{h \to 0} \frac{e^x e^h - e^x}{h}$$

$$= \lim_{h \to 0} \frac{e^x(e^h - 1)}{h}$$

$$f'(x) = e^x \lim_{h \to 0} \frac{e^h - 1}{h}$$

**b.** To find a formula for the derivative, you need to find $\lim\limits_{h \to 0} \dfrac{(e^h - 1)}{h}$.

Use your calculator or CAS to fill in the input–output tables given below. Find a reasonable value for the limit.

| $h$ | $\dfrac{e^h - 1}{h}$ |
|---|---|
| $-0.1$ | |
| $-0.01$ | |
| $-0.001$ | |
| $-0.0001$ | |
| $\downarrow -$ | $\downarrow$ |
| $\boxed{0}$ | $\boxed{\phantom{0}}$ |

| $h$ | $\dfrac{e^h - 1}{h}$ |
|---|---|
| $0.1$ | |
| $0.01$ | |
| $0.001$ | |
| $0.0001$ | |
| $\downarrow +$ | $\downarrow$ |
| $\boxed{0}$ | $\boxed{\phantom{0}}$ |

**c.** Using your estimate for $\lim\limits_{h \to 0} \dfrac{(e^h - 1)}{h}$, explain why it is reasonable to assume that $f'(x) = e^x$. That is, the derivative of $e^x$ <u>is</u> $e^x$ (which is why the number $e$ is so special!).

**3.** Find the derivative of $e^{f(x)}$ where $f$ is differentiable, by using the Chain Rule and the fact that the derivative of $e^x$ is $e^x$.

**a.** $e^{f(x)}$ can be expressed as the composition of two functions, $g$ and $f$. What is $g(x)$ in this case?

**b.** Use the Chain Rule to show that

$$\frac{d}{dx}(e^{f(x)}) = e^{f(x)} f'(x).$$

**c.** Thus, the derivative of $e^{f(x)}$ is found by multiplying $e^{f(x)}$ and $f'(x)$. Use this rule to show that

$$\frac{d}{dx}(e^{x^2 - 1}) = 2xe^{x^2 - 1}.$$

**4.** Find the derivative of the functions given below by using the rules for finding derivatives of combinations of functions along with the new rules.

$$\frac{d}{dx}(e^x) = e^x,$$

$$\frac{d}{dx}(e^{f(x)}) = e^{f(x)}f'(x)$$

**a.**   $y = 2e^{2x}$

**b.** $f(x) = x^3 e^{x+1}$

**c.** $g(t) = \sqrt{e^t}$

**d.** $h(r) = \dfrac{r^4 - 2}{4e^{r^2}}$

**e.** $y = e^{(x^3-1)^{100}}$

**f.** $f(\theta) = e^{\sin(\theta^2)}$

So, $e$ is a special number since the derivative of $e^x$ is $e^x$ and, consequently, the graph of $e^x$ and the graph of its derivative are the same. In the next task, you will find the derivative of the inverse of $e^x$ or $\ln(x)$ by converting $f(x) = \ln(x)$ to exponential form and then applying the formulas you discovered in the last task. To help convince yourself that the outcome is reasonable, you will analyze the behavior of the tangent line as it travels along the graph of $\ln(x)$. Finally, you will use the Chain Rule to find the derivative of $\ln(f(x))$ and practice using the new rules.

## Task 6-15: Finding the Derivatives of ln(x) and ln(f(x))

1. Show that

$$\text{if } f(x) = \ln(x), \text{ then } f'(x) = \frac{1}{x}$$

by completing the steps in the following proof. To help you along, we have started Step (2).

   **Step (1)** Convert $f(x) = \ln(x)$ to exponential form, noting that $f(x) = \log_e x$.

   **Step (2)** Take the derivative of *both* sides of the exponential form:

$$\frac{d}{dx}(e^{f(x)}) = \frac{d}{dx}(x)$$

$$=$$

   **Step (3)** Solve the equation in Step (2) for $f'(x)$.

   **Step (4)** According to Step (1), $e^{f(x)} = x$. Substitute $x$ for $e^{f(x)}$ in the expression for $f'(x)$. This completes the proof.

2. In part 1, you showed that the derivative of $\ln(x)$ is $\frac{1}{x}$. Convince yourself that this is reasonable by analyzing the behavior of the tangent line to the graph of $f(x) = \ln(x)$ for $x > 0$.

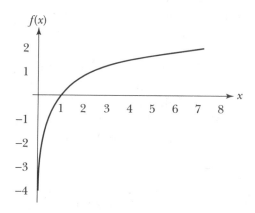

**a.** Describe the shape and location of the graph of the derivative of $f(x) = \ln(x)$ by analyzing the shape of the graph of $f$. Include in your description anything you can say about:

- values of $x$ where the graph of the derivative intersects the horizontal axis
- the location of the graph of the derivative with respect to the horizontal axis
- intervals where the derivative is increasing or decreasing
- the existence and location of turning points of the graph of the derivative

**b.** Analyze the behavior of the slope of the tangent line as it travels along the graph of $f(x) = \ln(x)$.

**(1)** Suppose the tangent line moves along the graph from left to right. What happens to the value of the slope of the tangent line as $x$ gets larger and larger? In other words, describe what is happening to the value of $f'(x)$ as $x \to +\infty$.

**(2)** Similarly, suppose the tangent line travels from right to left. What happens to the value of the slope of the tangent line as $x$ gets smaller and smaller? In other words, describe what is happening to the value of $f'(x)$ as $x \to 0^+$.

**c.** Since $\dfrac{d}{dx}(\ln(x)) = 1/x$, all the observations that you made in parts a and b about the shape, location, and behavior of the graph of the derivative of $\ln(x)$ should also hold for the graph of $y = 1/x$, for $x > 0$.

Sketch a graph of $y = 1/x$, for $x > 0$. Does the graph support all the observations you made above? If not, check where you went astray when you made your observations.

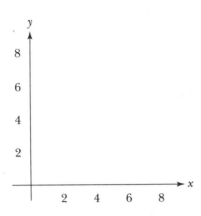

3. Find the derivative of $\ln(f(x))$ where $f$ is positive and differentiable by using the Chain Rule and the fact that the derivative of $\ln(x)$ is $1/x$.

   **a.** $\ln(f(x))$ can be expressed as the composition of two functions, $g$ and $f$. What is $g(x)$ in this case?

   **b.** Use the Chain Rule to show that

   $$\frac{d}{dx}(\ln(f(x))) = \frac{1}{f(x)}\, f'(x) = \frac{f'(x)}{f(x)}.$$

   **c.** Thus, the derivative of $\ln(f(x))$ is found by multiplying $\dfrac{1}{f(x)}$ and $f'(x)$. Use this rule to show that

   $$\frac{d}{dx}(\ln(x^2)) = \frac{2}{x}.$$

4. Find the derivative of the functions given below by using the rules for finding derivatives of combinations of functions and the new rules.

$$\frac{d}{dx}(\ln(x)) = \frac{1}{x},$$

$$\frac{d}{dx}(\ln(f(x))) = \frac{f'(x)}{f(x)}$$

a. $f(x) = \ln(2x^3) + 4x$

b. $f(u) = u\ln(u)$

c. $y = 2\,\dfrac{x}{\ln(x^2)}$

d. $h(t) = \ln(e^{t-1})$

e. $g(x) = (\ln(x^4 + x^2))^{100}$

f. $f(x) = \ln(\sin(2x))$

At this point, you can find the derivatives of $e^x$ and $\log_e x$ or $\ln x$, but what if the base is not equal to $e$? In the next task, you will find the derivatives of general exponential and logarithmic functions $a^x$ and $\log_a x$ by expressing these functions in terms of $e$ and ln and then applying the rules you developed in the last two tasks. Finally, as you have done several times before, you will use the Chain Rule to find formulas for the derivatives of $a^{f(x)}$ and $\log_a(f(x))$.

One comment before you begin. To express $a^x$ and $\log_a x$ in terms of $e$ and ln, you will need to use the following properties, where $r$ and $c$ are real numbers with $r > 0$:

- $\log_a r^c = c \log_a r$

- $a^{\log_a r} = r$

or when $a = e$:

- $\ln r^c = c \ln r$

- $e^{\ln r} = r$

Keep these two properties in mind as you do the next task.

## Task 6-16: Finding the Derivatives of General Exponential and Logarithmic Functions

1. Consider the general exponential functions.

   a. Show that for any positive number $a$,

   $$\frac{d}{dx}(a^x) = (\ln a)\,a^x.$$

   (1) The general exponential function $a^x$ can be expressed in terms of $e$ and ln, as illustrated in the steps given below. Provide an explanation for each step.

   | Step | Explanation |
   |------|-------------|
   | $y = e^{\ln y}$ | |
   | $a^x = e^{\ln a^x}$ | |
   | $a^x = e^{x \ln a}$ | |

   (2) Based on part (1), you know that $a^x = e^{x \ln a}$. Moreover, the steps listed below show that $\frac{d}{dx}(a^x) = (\ln a)\,a^x$. Provide an explanation for each step, noting that $\ln a$ is a constant, since $a$ is fixed.

   | Step | Explanation |
   |------|-------------|
   | $\frac{d}{dx}(a^x) = \frac{d}{dx}(e^{x \ln a})$ | |
   | $= e^{x \ln a}\frac{d}{dx}(x \ln a)$ | |
   | $= e^{x \ln a}\ln a$ | |
   | $\frac{d}{dx}(a^x) = (\ln a)\,a^x$ | |

**(3)** Use the new rule to show that $\dfrac{d}{dx}(3^x) = (\ln 3)3^x$.

**b.** Find the derivative of $a^{f(x)}$ where $f$ is differentiable by using the Chain Rule and the fact that the derivative of $a^x$ is $(\ln a)a^x$.

**(1)** $a^{f(x)}$ can be expressed as the composition of two functions, $g$ and $f$. What is $g(x)$ in this case?

**(2)** Use the Chain Rule to show that

$$\frac{d}{dx}(a^{f(x)}) = (\ln a)\, a^{f(x)} f'(x).$$

**(3)** Thus, the derivative of $a^{f(x)}$ equals the product of $\ln a$, $a^{f(x)}$, and $f'(x)$. Use this rule to show that

$$\frac{d}{dx}(3^{x^2}) = (\ln 3)3^{x^2}(2x).$$

**c.** Find the derivative of the functions given below by using the rules for finding derivatives of combinations of functions and the new rules for differentiating general exponential functions.

$$\frac{d}{dx}(a^x) = (\ln a)\, a^x,$$

$$\frac{d}{dx}(a^{f(x)}) = (\ln a)\, a^{f(x)} f'(x)$$

**(1)** $y = 8.5^x$

**(2)**  $h(t) = 5^{\sqrt{t}}$

**(3)**  $g(r) = 10^{r^2+1} + (r^2 + 1)^{10}$

**(4)**  $f(x) = x4^x$

**2.** Consider the general logarithmic functions.

**a.** Show that for any real positive number $a$,

$$\frac{d}{dx}(\log_a x) = \left(\frac{1}{\ln a}\right)\frac{1}{x} \quad \text{for } x > 0.$$

**(1)** Express the general logarithmic function $\log_a x$ in terms of ln by giving an explanation for each step listed below.

| Step | Explanation |
|---|---|
| $y = \log_a x$ | Let $y = \log_a x$. |
| $a^y = x$ | |
| $\ln a^y = \ln x$ | |
| $y \ln a = \ln x$ | |
| $y = \dfrac{\ln x}{\ln a}$ | |
| $\log_a x = \dfrac{\ln x}{\ln a}$ | |

**(2)** Based on part (1), you know that $\log_a x = \dfrac{\ln x}{\ln a}$ when $x > 0$. Next, show why $\dfrac{d}{dx}(\log_a x) = \left(\dfrac{1}{\ln a}\right)\dfrac{1}{x}$ by giving an explanation for each step listed below.

| Step | Explanation |
|---|---|
| $\dfrac{d}{dx}(\log_a x) = \dfrac{d}{dx}\left(\dfrac{\ln x}{\ln a}\right)$ | |
| $= \left(\dfrac{1}{\ln a}\right)\dfrac{d}{dx}(\ln x)$ | |
| $\dfrac{d}{dx}(\log_a x) = \left(\dfrac{1}{\ln a}\right)\dfrac{1}{x}$ | |

**(3)** Use the new rule to show that $\dfrac{d}{dx}(\log_3 x) = \left(\dfrac{1}{\ln 3}\right)\dfrac{1}{x}$.

**b.** Find the derivative of $\log_a f(x)$ where $f$ is differentiable by using the Chain Rule and the fact that the derivative of $\log_a x$ is $\left(\dfrac{1}{\ln a}\right)\dfrac{1}{x}$.

**(1)** $\log_a f(x)$ can be expressed as the composition of two functions, $g$ and $f$. What is $g(x)$ in this case?

**(2)** Use the Chain Rule to show that
$$\frac{d}{dx}(\log_a f(x)) = \left(\frac{1}{\ln a}\right)\frac{f'(x)}{f(x)} \quad \text{when } f(x) > 0.$$

**(3)** Thus, the derivative of $\log_a f(x)$ equals $f'(x)$ divided by the product of $f(x)$ and $\ln a$. Use this rule to show that
$$\frac{d}{dx}(\log_3 x^2) = \frac{2}{x \ln 3}.$$

**c.** Find the derivative of the functions given below by using the rules for finding derivatives of combinations of functions and the new rules.

$$\frac{d}{dx}(\log_a x) = \left(\frac{1}{\ln a}\right)\frac{1}{x} \qquad \text{for } x > 0$$

$$\frac{d}{dx}(\log_a f(x)) = \left(\frac{1}{\ln a}\right)\frac{f'(x)}{f(x)} \qquad \text{for } f(x) > 0$$

**(1)** $y = \log_{1/2} x$

**(2)** $h(t) = \log_5 \sqrt{t}$

**(3)** $g(r) = \dfrac{r^{10}}{\log_{10} r}$

**(4)** $f(x) = x \log_4 x$

Using the rules to find derivatives is much, much easier than using the definition of derivative. Using your CAS is even easier.

---

## Task 6-17: Evaluating Derivatives Using Your CAS

**1.** Use your CAS to find the derivative of some functions that you considered in the past two tasks using pencil and paper. Show that your CAS result is equivalent to your pencil and paper result.

**a.** $g(r) = \left( \dfrac{1}{r} + \dfrac{1}{r^2} \right)^{10}$

**(1)** What is the derivative using pencil and paper? (See Task 6-4, part 2c.)

**(2)** Find the derivative using your CAS.

**(3)** Show that the two results are equivalent.

**b.** $x(t) = \sqrt{t}(2t + 4)$

    **(1)** What is the derivative using pencil and paper? (See Task 6-5, part 2b.)

    **(2)** Find the derivative using your CAS.

    **(3)** Show that the two results are equivalent.

**c.** $y(p) = \dfrac{p}{4p^4 + 3p^3 + 2p^2 + p}$

    **(1)** What is the derivative using pencil and paper? (See Task 6-6, part 2c.)

    **(2)** Find the derivative using your CAS.

    **(3)** Show that the two results are equivalent.

**d.** $y = \dfrac{\sin(z^6)}{z^6 - \cos(z)}$

    **(1)** What is the derivative using pencil and paper? (See Task 6-13, part 2e.)

    **(2)** Find the derivative using your CAS.

    **(3)** Show that the two results are equivalent.

**e.** $y = e^{(x^3 - 1)^{100}}$

    **(1)** What is the derivative using pencil and paper? (See Task 6-14, part 4e.)

    **(2)** Find the derivative using your CAS.

    **(3)** Show that the two results are equivalent.

**f.** $g(x) = (\ln(x^4 + x^2))^{100}$

    **(1)** What is the derivative using pencil and paper? (See Task 6-15, part 4e.)

    **(2)** Find the derivative using your CAS.

    **(3)** Show that the two results are equivalent.

**2.** Consider $f(x) = 4x^3 - 12x^2 - 15x + 18.5$.

    **a.** Use your CAS to find the $(x, y)$-coordinates of all the horizontal tangents to the graph of $f$. List the points below.

    **b.** Use your CAS and pencil and paper to find the equation of the tangent line to the graph of $f$ at $x = 2$. Show your work in the space below.

**c.** Use your CAS to plot $f$ for $-2 \le x \le 4$ and the tangent line to the graph at $x = 2$ on the same pair of axes. Print your graph and place it below. On your printout label:

**(1)** The tangent line to the graph at $x = 2$

**(2)** The intervals where $f$ is increasing and decreasing

**(3)** The local extrema of $f$

## Summary of the Differentiation Rules, Part 2

*Note: We have listed all the rules that you examined in this section. However, you only need the four rules marked with an asterisk (\*), since the other rules follow from these.*

**Trigonometric functions**

$$\frac{d}{dx}(\sin(x)) = \cos(x)$$

* $$\frac{d}{dx}(\sin(f(x))) = \cos(f(x)) \cdot f'(x)$$

$$\frac{d}{dx}(\cos(x)) = -\sin(x)$$

* $$\frac{d}{dx}(\cos(f(x))) = -\sin(f(x)) \cdot f'(x)$$

**Exponential and natural logarithmic functions**

$$\frac{d}{dx}(e^x) = e^x$$

$$\frac{d}{dx}(e^{f(x)}) = e^{f(x)} f'(x)$$

$$\frac{d}{dx}(\ln(x)) = \frac{1}{x} \quad \text{for } x > 0$$

$$\frac{d}{dx}(\ln(f(x))) = \frac{f'(x)}{f(x)} \quad \text{for } f(x) > 0$$

**General exponential and logarithmic functions ($a > 0$)**

$$\frac{d}{dx}(a^x) = (\ln a) a^x$$

* $$\frac{d}{dx}(a^{f(x)}) = (\ln a) a^{f(x)} f'(x)$$

$$\frac{d}{dx}(\log_a x) = \left(\frac{1}{\ln a}\right)\frac{1}{x} \quad \text{for } x > 0$$

* $$\frac{d}{dx}(\log_a f(x)) = \left(\frac{1}{\ln a}\right)\frac{f'(x)}{f(x)} \quad \text{for } f(x) > 0$$

## Unit 6 Homework After Section 3

- Complete the tasks in Section 3. Be prepared to discuss them in class.

- Evaluate some derivatives in HW6.15.

**HW6.15** Find the derivative of each of the following functions.

**1.** $h(\alpha) = \cos^9(\alpha)$

**2.** $y = e^{e^x}$

**3.** $f(t) = 6\sin(t)\sin(6t)$

**4.** $g(t) = \ln\sqrt{e^{2t} + e^{-2t}}$

**5.** $f(p) = \dfrac{\sin^8(8p)}{\cos^6(6p)}$

**6.** $f(x) = \dfrac{\ln(e^x + 1)}{\ln(e^x - 1)}$

**7.** $f(\theta) = \dfrac{\theta^3 - 2\theta^2 + \theta}{\cos(2\theta + 1)}$

**8.** $y = \dfrac{x}{6^x + x^6}$

**9.** $f(t) = e^{\sqrt{t}} + \sqrt{e^t}$

**10.** $g(r) = \dfrac{\log_2 r}{\log_3 r}$

**11.** $f(x) = \sin(\ln(x))$

**12.** $y = \ln(\log_{10} x)$

**13.** $g(t) = \ln(\cos^2(t))$

**14.** $f(x) = 2^{2-x^2}$

**15.** $h(r) = e^{\ln(r^2)}$

**16.** $y = 3^{\sqrt{x}}$

**17.** $y = \cos(e^{x^2})$

**18.** $f(x) = \log_8 \sqrt[3]{(2x + 5)^2}$

**19.** $f(x) = e^{\sin(x) + \cos(x)}$

**20.** $g(r) = (r^2 + 1)^{10} \log_{10}(r^2 + 1)$

- Find the derivatives of the other basic trigonometric functions in HW6.16.

**HW6.16**

**1.** Find the rules for differentiating the other basic trigonometric functions. Show that:

**a.** $\dfrac{d}{dx}(\csc(x)) = -\csc(x)\cot(x)$

**b.** $\dfrac{d}{dx}(\sec(x)) = \sec(x)\tan(x)$

**c.** $\dfrac{d}{dx}(\tan(x)) = \sec^2(x)$

**d.** $\dfrac{d}{dx}(\cot(x)) = -\csc^2(x)$

*Hint: Recall that csc(x), sec(x), tan(x), and cot(x) can be expressed in terms of sin(x) and cos(x), where the following identities hold:*

$$\csc(x) = \frac{1}{\sin(x)}, \qquad \sec(x) = \frac{1}{\cos(x)},$$

$$\tan(x) = \frac{\sin(x)}{\cos(x)}, \qquad \cot(x) = \frac{\cos(x)}{\sin(x)},$$

  **i.** Find the derivative of the right side of each identity by using the Quotient Rule and the rules for differentiating $\sin(x)$ and $\cos(x)$.

  **ii.** Simplify the result by utilizing the identities and the fact that $\sin^2(x) + \cos^2(x) = 1$.

2. As usual, if $f$ is a differentiable function, then the Chain Rule and the rules for differentiating the basic functions which you proved in part 1 give the following more general rules:

$$\frac{d}{dx}(\csc(f(x))) = -\csc(f(x))\cot(f(x))f'(x),$$

$$\frac{d}{dx}(\sec(f(x))) = \sec(f(x))\tan(f(x))f'(x),$$

$$\frac{d}{dx}(\tan(f(x))) = \sec^2(f(x))f'(x),$$

$$\frac{d}{dx}(\cot(f(x))) = -\csc^2(f(x))f'(x).$$

Use these new rules, in conjunction with the ones you already know, to find the derivatives of the following functions.

**a.** $f(x) = \sec(x) - \sqrt{2}\tan(x)$

**b.** $g(t) = \dfrac{\tan(t)}{\cot(t) + 1}$

**c.** $y = \dfrac{2(x^2 + 1)\tan(x)}{3x - \csc(x)}$

**d.** $h(\theta) = \csc(4\theta^2) - \sin(e^\theta)$

**e.** $j(r) = 5r^3 \cot^5(3r)$

**f.** $k(t) = (t^3 - \cot(t^3 - t^2))^{-2}$

**g.** $p(y) = 4y^4 \sec\left(\dfrac{1}{y}\right)$

**h.** $s(t) = \sin(\tan(2t))$

**i.** $q = \sqrt{h}\cot^3(\sqrt{h})$

**j.** $l(z) = \dfrac{\tan(2z^2 + 1)}{\sec(4z^5 - z)}$

• Consider some applications in HW6.17 to HW6.19.

**HW6.17** The following function gives the layout of the tracks for a roller coaster, where $x$ meters from the starting point the track is $y$ meters above the ground:

$$y = 15 + 15\sin\left(\tfrac{\pi}{50}x\right).$$

1. Use your CAS to graph the layout of the roller coaster, assuming $0 \le x \le 100$ meters. Print a copy of your graph.

2. Find the distance above the ground (in meters) at the starting point of the track.

3. Find the distance above the ground (in meters) at the terminal point of the track.

4. Find the values of $x$ where the track has a turning point—that is, where it turns up or down.

5. Find the value of $x$ where the height of the track is an absolute maximum.

6. Find the value of $x$ where the height of the track is an absolute minimum.

**HW6.18** Your company's total cost, in millions of dollars, is given by

$$C(t) = 200 - 40e^{-t},$$

where $t$ is time.

**1.** Find the marginal cost of your company, $C'(t)$.

**2.** Find your marginal costs initially, that is, when $t = 0$.

**3.** Show that your company's total cost is always increasing as time passes.

**HW6.19** Since you enjoy taking exams (and since the Psychology Depart-ment was offering good pay), you signed up, along with other members of your anthropology course, for a psychology experiment whose purpose was to measure how much you remember . . . or how much you forget. After taking the anthropology final, the participants in the experiment took equivalent forms of the final at monthly intervals thereafter. The average score $S(t)$ in percent, after $t$ months, was found to be given by

$$S(t) = 78 - 15\ln(t + 1), \quad t \geq 0.$$

**1.** What was the average score when you initially took the test, that is, when $t = 0$?

**2.** What was the average score after 4 months?

**3.** What was the average score after 24 months?

**4.** What percentage of the initial score did people retain after two years (24 months)?

**5.** Find $S'(t)$. What does this tell you about $S(t)$?

**6.** Find the absolute maximum and minimum average scores and the times at which they occurred.

• Use your CAS to analyze some functions in HW6.20.

**HW6.20** Use your CAS to do the following exercises.

**1.** For the functions given below in a–c:

   **i.** Determine the intervals where the function is increasing/decreasing by (1) finding the zeros of the first derivative and then (2) evaluating the derivative at points between the zeros.

   **ii.** Find the coordinates of all the local extrema of the function.

   **iii.** Determine the intervals where the function is concave up/down by (1) finding the zeros of the second derivative and then (2) evaluating the second derivative at points between the zeros.

   **iv.** Find the coordinates of all the inflection points of the function.

v. Plot the function over an interval that includes all the local extrema and inflection points.

vi. Print a copy of your computer file.

- Mark your answers to the above questions on your printout.
- Label the intervals where the function is increasing/decreasing and concave up/down, and the local extrema and inflection points on your graph.

a. $f(x) = x^4 - 8x^2 + 16$

b. $g(x) = 9x - x^3$

c. $h(x) = x^2 - 3x + 2$

2. Find the absolute extrema of the following functions on the indicated interval by comparing the values of the function at the local extrema located in the interval and the value at each of the end points of the interval.

a. $f(x) = x^2 - 3x + 1$  on $[-1, 2]$

b. $f(x) = x^3 - 3x + 2$  on $[0, 2]$

- Extend the differentiation rules for logarithms in HW6.21.

## HW6.21

1. Consider

$$\ln|x| = \begin{cases} \ln x & \text{if } x > 0 \\ \ln(-x) & \text{if } x < 0. \end{cases}$$

a. Sketch a graph of $y = \ln|x|$, where $x \neq 0$.

b. Show that $\dfrac{d}{dx}(\ln|x|) = \dfrac{1}{x}$ for all $x \neq 0$.

*Hint: Using the rules you already know, $\dfrac{d}{dx}(\ln x) = 1/x$ for $x > 0$. Therefore, you only need to show that $\dfrac{d}{dx}(\ln(-x)) = 1/x$ for $x < 0$. To do this, consider $f(x) = -x$ for $x < 0$ and use the current version of the rules.*

2. Similarly,

$$\ln|f(x)| = \begin{cases} \ln(f(x)) & \text{if } f(x) > 0 \\ \ln(-f(x)) & \text{if } f(x) < 0. \end{cases}$$

a. Show that $\dfrac{d}{dx}(\ln|f(x)|) = \dfrac{f'(x)}{f(x)}$ for $f(x) \neq 0$.

*Hint: Express $\ln|f(x)|$ as the composition of two functions. Find the derivative by applying the Chain Rule and using the fact that $\dfrac{d}{dx}(\ln|x|) = 1/x$ for all $x \neq 0$.*

**b.** Use the new rule to find the derivatives of the following functions:

**(1)** $g(t) = \ln\left|\frac{1}{t}\right|$

**(3)** $f(x) = \ln\left|x^3 - 3\right|$

**(2)** $h(r) = \dfrac{1}{\ln|2r - 15|}$

**(4)** $y = (\ln|1 - e^{2x}|)^3$

- Prove two rules, one old and one new, in HW6.22.

**HW6.22**

1. In HW6.5, you proved the Power Rule for positive integers by considering the definition of derivative and using the Binomial Theorem. Now prove the Power Rule for any real exponent by applying the properties of logarithmic and exponential functions and the rules for finding their derivatives. That is, show that

$$\frac{d}{dx}(x^c) = cx^{c-1} \quad \text{when } c \text{ is any real number.}$$

*Hint: Based on the properties of exponential and logarithmic functions, you know that*

$$x^c = e^{\ln(x^c)} = e^{c\,\ln(x)}.$$

*Use the rules for differentiating exponential and logarithmic functions to show that*

$$\frac{d}{dx}(e^{c\,\ln(x)}) = cx^{c-1}.$$

2. Observe that you cannot use the Power Rule to find the derivative of $f(x) = x^x$, since the exponent is not a fixed real number. Show that

$$\frac{d}{dx}(x^x) = x^x(1 + \ln(x)) \quad \text{for } x > 0.$$

*Hint: Similar to the proof in part 1, first observe that*
$$x^x = e^{\ln(x^x)} = e^{x\,\ln(x)}.$$

*Now use the rules for differentiating exponential and logarithmic functions to show that*

$$\frac{d}{dx}(e^{x\,\ln(x)}) = x^x(1 + \ln(x)).$$

- Consider the probability density function in HW6.23.

**HW6.23** If a random variable is normally distributed with mean $\mu$ and standard deviation $\sigma$, then the *probability density function* is given by

$$f(x) = \frac{1}{\sigma\sqrt{2\pi}}\, e^{-\frac{1}{2}\left(\frac{x-\mu}{\sigma}\right)^2}.$$

This is an important function in statistics.

1. Use your CAS to graph of the probability density function for various values of $\mu$ and $\sigma$. For example, graph $f$ when $\mu = 75$ and $\sigma = 10$, when $\mu = 85$ and $\sigma = 10$, and when $\mu = 65$ and $\sigma = 15$.

    **a.** Print copies of the graphs of $f$ for several values of $\mu$ and $\sigma$.

    **b.** On each of the graphs mark the locations of the local extrema of $f$ and the inflection points of $f$. In each case, estimate the $x$-coordinates of the local extrema and the inflection points.

**2.** Describe the general shape of the graph of the probability density function.

**3.** Use calculus to show that, in general, the probability density function has a local maximum at $x = \mu$.

**4.** Use calculus to show that, in general, the probability density function has inflection points at $x = \mu + \sigma$ and at $x = \mu - \sigma$.

- Write your journal entry for this unit. As usual, before you begin to write, review the material in the unit. Think about how it all fits together. Try to identify what, if anything, is still causing you trouble.

**HW6.24** Write your journal entry for Unit 6.

**1.** Reflect on what you have learned in this unit. Describe in your own words the concepts that you studied and what you learned about them. How do they fit together? What concepts were easy? Hard? What were the important ideas? Give some examples of the main ideas.

**2.** Reflect on the learning environment for the course. Describe the aspects of this unit and the learning environment that helped you understand the concepts you studied. What activities did you like? Dislike?

# *Unit 7:*

# DEFINITE INTEGRALS: THE CALCULUS APPROACH

*The ideas of calculus were "in the air" in the 1650s and 1660s. The ingenious area calcu-lations and tangent constructions of Cavalieri, Torricelli, Fermat, Pascal, Barrow, and oth-ers were so suggestive that the final discovery of calculus as an autonomous discipline was almost inevitable within a very few years. The last steps of putting it all together were taken by two men of great genius working independently of each other: by Isaac Newton in what he called "the plague years of 1665 and 1666," and also by Gottfried Wilhelm Leibniz during his sojourn in Paris from 1672 to 1676.*

George F. Simmons, *Calculus Gems:*
*Brief Lives of Memorable Mathematicians,*
McGraw-Hill, 1992, p. 141.

## OBJECTIVES

1. Examine other situations where the Riemann sum approach applies.

2. Given a function *f*, represent the *accumulation function* generated by *f* in terms of a definite integral. Approximate values of the accumulation function using *partial Riemann sums*.

3. Given a function *f*, find an *antiderivative* of *f*. Explore the relationships between the antiderivatives of *f*. Discover rules for finding the anti-derivatives of basic functions.

4. Develop a conceptual understanding of the Fundamental Theorem of Calculus by examining the link between the accumulation function gen-erated by a function and the antiderivatives of the function.

5. Use the Fundamental Theorem of Calculus to evaluate definite integrals.

## OVERVIEW

In Unit 5, you developed a general method for approximating the area under a curve using the Riemann sum approach, and in the process, you discovered a definition for the definite integral. In particular, you considered a continuous, non-negative function $f$ defined over a closed interval $[a, b]$. You let

$$\Delta x = \frac{b - a}{n} \quad \text{and} \quad x_i = a + i\Delta x \quad \text{for } 0 \le i \le n,$$

where $n$ is a positive integer.

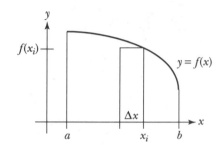

You discovered the following:

• The area of the $i$th rectangle equals $f(x_i)\Delta x$.

• The approximate area of the region under the graph of $f$ and over the interval $[a, b]$ is given by the Riemann sum $\Sigma_{i=1}^n f(x_i)\Delta x$.

• The exact area of the region under $f$ and over $[a, b]$ equals $\lim_{n \to \infty} \Sigma_{i=1}^n f(x_i)\Delta x$ or $\lim_{\Delta x \to 0} \Sigma_{i=1}^n f(x_i)\Delta x$ and is represented by the definite integral $\int_a^b f(x)dx$.

In this unit, you will examine some more situations where the Riemann sum approach applies. In particular, you will look at ways to approximate the following:

• The accumulated distance when the velocity varies
• The work done when the force varies
• The mass of an object when its density varies
• The area between two curves, where the distance between the curves varies

You will also consider how the sign of a function affects the Riemann sum and the associated definite integral. If the graph of a function dips below the horizontal axis, how does this affect the sign of the definite integral? Under what conditions is a definite integral positive? Negative? Zero?

Completing the tasks in the first section will help you develop a firmer understanding of what a definite integral is and how the Riemann sum approach applies to other situations. But, unless a definite integral corresponds to a regularly shaped region, the only way you can find the exact value of the definite integral is by evaluating the limit of the associated Riemann sum. Taking the limit of a Riemann sum can be a daunting task. There must be an easier and more direct way to do this. And there is. Finding this more straightforward way involves examining the relationship between two functions that are associated with a given function. You are familiar with one of these functions; the other is new.

**1.** The first function is the *accumulation function* generated by a given function *f*, which you first considered in Task 5-13.

For instance, if *f* is a non-negative continuous function, defined on the closed interval [*a*, *b*], you can find the area of the region bounded by *f* by taking the limit of the Riemann sum, which is represented by the definite integral $\int_a^b f(t)dt$. Similarly, if *x* is *any* value between *a* and *b*, you can find the area under *f* and over the interval [*a*, *x*] by taking the limit of the corresponding *partial* Riemann sum, which is represented by the definite integral $\int_a^x f(t)dt$. Note that the bounds on this new integral go from *a* to *x*, not *a* to *b*.

This process defines a new function.

$$A(x) = \int_a^x f(t)\ dt \quad \text{for } a \le x \le b.$$

*A* is called the *accumulation function defined by f*, since when you give *A* a value of *x* between *a* and *b*, *A* returns the accumulated area under the graph of *f* from *a* to *x*.

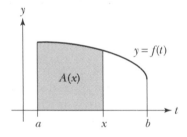

**2.** The other function you will examine is an *antiderivative* of a given function *f*.

The question here is: Given a function *f*, what function can you differentiate to get *f*? That is, how can you reverse the process of differentiation? For instance, if

$$f(x) = 2x,$$

what function can you differentiate to get 2*x*? $F(x) = x^2$ is one possibility, as is $G(x) = x^2 - \sqrt{2}$. In fact, there are infinitely many functions whose derivative is 2*x*. Can you find another one?

On the surface these two ideas—(1) representing an accumulation by a function and (2) reversing the process of differentiation—seem to have very little in common, but it turns out that they are related in an extremely important way:

The accumulation function generated by a given function is an antiderivative of the given function.

Another way of saying this is

> The derivative of the accumulation function generated by a given function equals the given function.

This important connection is the foundation of the Fundamental Theorem of Calculus. This is not a new idea. At the end of Unit 5, in Task 5-13 and in HW5.13, you showed that the connection holds for linear functions. In this unit, you will show that it holds for any continuous function, and you will see how it provides a direct way to evaluate a definite integral without calculating the limit of a Riemann sum.

## SECTION 1

### Applying the Riemann Sum Approach to Other Situations

In Unit 5, you examined how to apply the Riemann sum approach to find the area of a region lying under the graph of a continuous function $f$ and over a closed interval $[a, b]$. You approximated the area of the region by dividing it up into smaller subregions and then approximating the area of each subregion. Some observations:

i. You approximated the area of each subregion—whose height was *varying*—by using the formula for calculating the area of a shape whose height is *constant*, namely

$$\text{area} = \text{base} \cdot \text{height}.$$

In other words, you used the area formula for a familiar shape, namely a rectangle, to approximate the area of each subregion.
ii. Then, you approximated the area of the entire region by summing the approximate areas of the subregions.
iii. Finally, you found the exact value of the area of the region by taking the limit of the sum of these approximations, and you expressed the result in terms of a definite integral.

This same approach—using a formula for the constant case to approximate the result when a quantity varies—can be applied to a number of other situations. For example, suppose over a specified time interval, you want to find the accumulated distance of an object when the velocity of the object is varying. Mimicking the approach you used in the area situation, you can divide the given time interval into smaller time intervals and then:

i. You can approximate the distance the object travels during each of the smaller time periods—where the velocity may be varying—by using the formula for calculating distance when the velocity is constant, namely

$$\text{distance} = \text{rate} \cdot \text{time}.$$

ii. Then, you can approximate the accumulated distance by summing the distances the object travels during each of the smaller time periods.

**iii.** Finally, you can find the exact value of the accumulated distance by taking the limit of the sum of these approximations, and you can express the result in terms of a definite integral.

You can use the same approach to calculate the amount of work it takes to move an object a given distance, when the force varies. In this case, you approximate the amount of work it takes to move the object over smaller distances by using the formula when the force is constant, namely

$$\text{work} = \text{force} \cdot \text{distance}.$$

Calculating the accumulated distance of an object when its velocity varies, or the amount of work done when the force varies, may sound quite different from calculating the area of a region whose height varies. The Riemann sum approach, however, is applicable to these situations (and to many others as well). In this section, you will examine more closely why this is true. As you do so, keep in mind the general approach:

- Divide the given interval into equal-sized subintervals.

- Approximate the desired quantity for each subinterval by using the formula for the constant case.

- Approximate the desired quantity for the entire interval by adding up the approximations for each subinterval.

- Find the exact value by taking the limit of the sum of the approximations.

- Express the result in terms of a definite integral.

---

## Task 7-1: Finding a Formula for Distance When the Velocity Varies

1. *Will Darcy get there on time?* Darcy is due at a meeting in 2 minutes. It is scheduled to take place on the other side of the campus, and, as usual, she is running late. She knows she can make it if she runs super fast without ever slowing down, but unfortunately she is not in top shape. Examine whether or not she makes it.

   **a.** Knowing that she is late and being somewhat frazzled, Darcy takes off and runs as fast as she can, moving at the rate of 5 meters per second (m/sec).

   (1) Find the distance Darcy runs between 0 and 10 seconds.

   (2) Find the distance Darcy runs between 0 and 20 seconds.

   **b.** At the end of 20 seconds, she begins to tire and slows her rate to 4.5 m/sec. She runs for the next 20 seconds at this new rate. Estimate the distance Darcy runs between 20 and 40 seconds.

**c.** Darcy becomes increasingly winded and continues to slow down every 20 seconds. In particular, after each successive 20-second time period she slows down to 3.8 m/sec, 3.4 m/sec, 2.8 m/sec, and 2.5 m/sec. Approximate the distance Darcy runs during each consecutive 20-second time period, and enter your results onto the table given below.

*Note: Darcy's "accumulated distance" or her distance-to-date is her distance at the end of the current time period from the starting point.*

| No. of time intervals | Right end-point of *i*th time interval | Velocity at end of *i*th time interval | Length of *i*th time interval | Approximate distance for *i*th time interval | Approximate accumulated distance |
|---|---|---|---|---|---|
| *i* | *t* (sec) | *v* (m/sec) | Δ*t* (sec) | *v(t)*Δ*t* (m) | *D*(m) |
| 1 | 20 | 5.0 | 20 | 100 | 100 |
| 2 | 40 | 4.5 | 20 | 90 | 190 |
| 3 | | | | | |
| 4 | | | | | |
| 5 | | | | | |
| 6 | | | | | |

**d.** If the room where Darcy's meeting is being held is 450 meters away from her staring point, does it appear that she makes it? Justify your conclusion.

**e.** Sketch Darcy's velocity function *v* on the axes given below. Make your graph realistic. In particular, make it reflect the facts that when she starts to run it takes a second or so for her to get up to speed, and that when she slows down at the end of each 20-second time interval, it takes a second or so to establish her new speed.

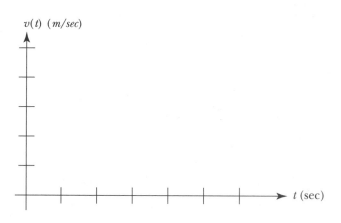

**f.** $v(60)$ m/sec $\times$ 20 sec approximates the distance that Darcy runs during the third 20-second time interval; that is, the distance she runs when $40 \leq t \leq 60$.

(1) Explain why this is the case.

(2) This distance corresponds to the area of a rectangular region on the graph of $v$. Shade the region on your graph in part e.

**g.** Assume that you estimate her velocity during a specific time interval by calculating her velocity at the right endpoint of the time interval. Then, just as $v(60)$ m/sec $\times$ 20 sec gives the approximate distance that Darcy runs between 40 and 60 seconds, each of the following expressions approximates the distance Darcy runs during a particular 20-second time interval. For each of the expressions:

**i.** State the time interval.

**ii.** Shade the associated rectangle on the velocity graph in part e.

(1) $v(20)$ m/sec $\times$ 20 sec

(2) $v(80)$ m/sec $\times$ 20 sec

(3) $v(120)$ m/sec $\times$ 20 sec

**2.** Consider the general case when the velocity is non-negative.

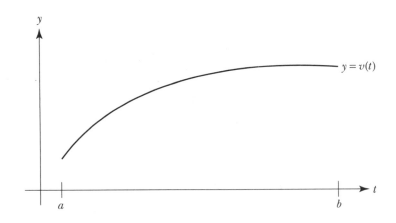

**a.** Suppose $y = v(t)$ gives the velocity of an object at time $t$, for $a \leq t \leq b$. Assume $v(t) \geq 0$. As usual, divide the interval $[a, b]$ into $n$ subintervals each of length $\Delta t$, where $t_i$ is the right endpoint of the $i$th subinterval, for $1 \leq i \leq n$. Suppose the graph of $v$ is given above. Label $t_1$, $t_2$, $t_{i-1}$, $t_i$, $t_{n-1}$, and $t_n$ on the $t$ axis.

**b.** When the velocity (rate) is constant, you can calculate the distance an object moves during a specified time period using the formula distance = rate · time. Use this fact to approximate the distance an object moves during a typical time interval, say the $i$th time interval, where $1 \leq i \leq n$, when the velocity is varying.

   (1) Approximate the velocity during the $i$th time interval by evaluating $v$ at the right endpoint of the subinterval.

   (2) Approximate the distance the object moves during the $i$th time interval by applying the formula for the constant case.

   (3) Sketch the associated rectangle on the graph of $v$, labeling the length of the subinterval $\Delta t$ and the approximate velocity during time interval $v(t_i)$.

**c.** Approximate the accumulated distance of the object between time $t = a$ and $t = b$ by summing the estimates from part b and using sigma notation.

**d.** Find a formula for the accumulated distance of the object by taking the appropriate limit of the sum in part c.

**e.** Use a definite integral to express the accumulated distance of the object between time $t = a$ and $t = b$, when $y = v(t)$.

**3.** Darcy stopped to chat after her meeting and now she has to get to class, pronto. Unfortunately, her professor takes roll at the beginning of the period, and if she is not there on time, she is marked absent. She "borrows" Bob's bike (which he had left leaning against the front of the building) and takes off, sighing to herself, "Why are my days always so hectic?" Her rate in feet per second is described by the function

$$v(t) = 6t - \frac{t^2}{5}.$$

**a.** Sketch a graph of Darcy's velocity function for $0 \leq t \leq 30$ seconds.

**b.** Suppose you divide the time interval [0, 30] into 15 equal subintervals. Also, suppose you estimate Darcy's velocity during a particular time subinterval by finding her velocity at the right endpoint of the subinterval. For each of the following time periods:

  **i.** Estimate her distance.

  **ii.** Shade the associated rectangle(s) on the graph in part a.

  **iii.** State whether the approximation is too large or too small.

**(1)** From 0 to 2 seconds

**(2)** From 0 and 4 seconds

**(3)** Between 14 and 16 seconds

**(4)** Between 28 and 30 seconds

**c.** Represent the exact distance Darcy peddles from $t = 0$ to $t = 30$ seconds by a definite integral.

The Riemann sum approach can be used to approximate the distance an object travels, as well as the area under a curve, and the exact values of these quantities can be expressed as definite integrals, where

- The accumulated area of the region under the graph of a function $f$ and over the closed interval $[a, b]$ is given by

$$\text{area} = \lim_{\Delta x \to 0} \sum_{i=1}^{n} \underbrace{f(x_i)\Delta x}_{\substack{\text{area of } i\text{th} \\ \text{rectangle}}} = \int_{a}^{b} f(x)\ dx.$$

approximate area of region under curve

exact area of region under the graph of $f$ over the interval $[a, b]$

- The accumulated distance of an object, from time $t = a$ to $t = b$, whose velocity is $v$ is given by

$$\text{distance} = \lim_{\Delta t \to 0} \sum_{i=1}^{n} \underbrace{v(t_i)\Delta t}_{\substack{\text{approximate} \\ \text{distance} \\ \text{during } i\text{th} \\ \text{time interval}}} = \int_{a}^{b} v(t)\ dt.$$

approximate accumulated distance

exact accumulated distance

Note the similarities between these two applications of a definite integral. In HW7.2 and HW7.3, you will use the same approach to examine some other situations.

Thus far, we have only considered functions whose graphs lie above the closed interval—that is, functions that are non-negative. In this case, the value of the associated definite integral is non-negative. But, what if the graph of a function dips below the horizontal axis? How does this affect the definite integral? In the next task, you will use the Riemann sum approach to examine how the sign of $f(x)$ affects the sign of the associated definite integral.

## Task 7-2: Integrating Functions Whose Graphs Dip Below the Axis

**1.** Consider a specific example.

**a.** On the axis below, sketch of the graph of $g(x) = -x + 1$ for $-1 \le x \le 2$.

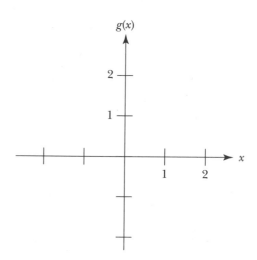

**b.** Approximate the definite integral $\int_{-1}^{2} g(x)\ dx$, where $g(x) = -x + 1$, using the Riemann sum approach. Suppose $n = 6$ and, as usual, let $x_i$ be the right endpoint of the $i$th subinterval for $1 \le i \le 6$. On the graph you drew in part a, carefully sketch the six rectangles that correspond to the components of the Riemann sum.

**c.** Calculate the Riemann sum where $n = 6$.

   **(1)** Find the value of $\Delta x$.

**(2)** Fill in the table.

| $i$ | $x_i$ | $g(x_i)$ | $g(x_i)\Delta x$ |
|---|---|---|---|
| 1 | | | |
| 2 | | | |
| 3 | | | |
| 4 | | | |
| 5 | | | |
| 6 | | | |

**(3)** Find the value of the Riemann sum.

$$\sum_{i=1}^{6} g(x_i)\Delta x =$$

**d.** Examine how the sign of $g(x)$ affects the value of the Riemann sum.

**(1)** List the values of $x_i$ where $g(x_i)$ is negative—that is, where the graph of $g$ dips below the horizontal axis. Describe how the value of $g(x_i)\Delta x$ contributes to the value of the Riemann sum in this case.

**(2)** Similarly, list the values of $x_i$ where $g(x_i)$ is zero and describe how the value of $g(x_i)\Delta x$ contributes to the value of the Riemann sum in this case.

**(3)** Finally, list the values of $x_i$ where $g(x_i)$ is positive and describe how the value of $g(x_i)\Delta x$ contributes to the value of the Riemann sum in this case.

**2.** Generalize your observations from part 1d by describing how the sign of a function affects the sign of the value of the associated definite integral. In particular, suppose $f$ is an arbitrary continuous function. For each of the following situations, find the sign (positive, negative, zero) of the value of $\int_a^b f(x)\ dx$. Justify your conclusion.

**a.** The graph of $f$ lies below the horizontal axis, for all $a \le x \le b$.

**b.** The graph of $f$ lies above the horizontal axis, for all $a \leq x \leq b$.

**c.** The graph of $f$ passes through the horizontal axis at some point between $a$ and $b$.

3. Consider $g(x) = -x + 1$ again. For each of the definite integrals listed below:

   **i.** Sketch a small graph of $g$ for $-1 \leq x \leq 2$ and shade the region represented by the definite integral.

   **ii.** Find the exact value of each definite integral, using a geometric approach.

   *Caution: Watch your signs! Keep in mind that the portion of the graph that lies below the horizontal axis makes a negative contribution to the Riemann sum and hence to the value of the definite integral.*

   **a.** $\int_{-1}^{1} g(x)\ dx$

   **b.** $\int_{1}^{2} g(x)\ dx$

   **c.** $\int_{-1}^{2} g(x)\ dx$

   **d.** Describe the relationship among the values of the three integrals in parts a, b, and c.

**4.** Suppose $f$ is a continuous function defined over the interval $[a, b]$. For each of the following cases:

**i.** Sketch several functions whose integrals have the indicated sign (positive, negative, or zero).

**ii.** Describe the general behavior of the graph of $f$.

**a.** Case 1: $\displaystyle\int_a^b f(x)\ dx > 0$

**b.** Case 2: $\displaystyle\int_a^b f(x)\ dx < 0$

**c.** Case 3: $\displaystyle\int_a^b f(x)\ dx = 0$

When the graph of a function dips below the axis, the portion below the axis contributes a negative amount to the associated definite integral. If the region above the axis is larger than the region below, the definite integral is positive; if the region below is larger than the region above, the definite integral is negative; if the two regions are equal, the definite integral is 0. Keep this in mind as you do the following task.

## Task 7-3: Interpreting Definite Integrals (Again)

**1.** Consider the graph of *h*.

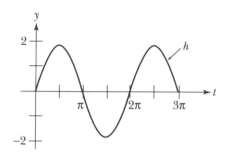

For each of the following integrals:

  **i.** Use different patterns to shade in the region on the graph associated with the definite integral.

  **ii.** Determine if the value of the integral is negative, zero, or positive. Justify your response.

  **a.** $\displaystyle\int_{\pi/2}^{3\pi/2} h(t)\ dt$

  **b.** $\displaystyle\int_{0}^{3\pi} h(t)\ dt$

  **c.** $\displaystyle\int_{\pi+0.01}^{2\pi-0.01} h(t)\ dt$

**2.** Consider the graph of *f*.

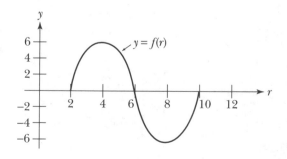

Determine whether each of the following statements is true or false. *Justify your response.*

**a.** $\displaystyle\int_{2}^{6} f(r)\,dr \approx -\int_{6}^{10} f(r)\,dr$

**b.** $\displaystyle\int_{2}^{6} f(r)\,dr + \int_{6}^{10} f(r)\,dr = \int_{2}^{10} f(r)\,dr$

**c.** $\displaystyle\int_{2}^{6} f(r)\,dr < 24$

**d.** $\displaystyle\int_{4}^{6} f(r)\,dr > \frac{1}{2}f(4.5) + \frac{1}{2}f(5.0) + \frac{1}{2}f(5.5) + \frac{1}{2}f(6.0)$

Hint: Each term on the right side of the inequality represents the area of a rectangle.

**e.** $\displaystyle\int_{4}^{8} f(r)\,dr = 0$

3. Recall that the velocity function is the first derivative of the distance function and describes the rate of change of an object. Moreover, the accumulated distance of the object between time $t = a$ and $t = b$ corresponds to the area of the region bounded by the velocity graph, where the accumulated distance equals

$$\int_{a}^{b} v(t)\,dt.$$

When the velocity function is positive (as it was in Task 7-1), then:

- The object is moving forward.
- The corresponding distance function is increasing.
- The graph of the velocity function lies above the horizontal axis.
- The velocity function is contributing a *positive* amount to the accumulated distance.

However, when the velocity is negative, then:

- The object is moving backward.
- The corresponding distance function is decreasing.
- The graph of the velocity function lies below the horizontal axis.
- The velocity function is contributing a *negative* amount to the accumulated distance.

**a.** For each of the following velocity graphs, describe the movement of the object with respect to its starting point and give the sign of the accumulated distance. In particular:

**i.** Give the time interval(s) when the object is moving forward.

**ii.** Give the time interval(s) when the object is moving backward.

**iii.** Determine whether the accumulated distance—the distance from the starting point at time $t = b$—is positive, negative, or zero by considering the area of the region bounded by the graph.

**(1)** $v(t)$

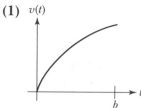

**(2)** $v(t)$

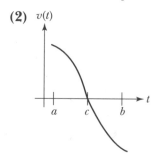

**(3)** $v(t)$

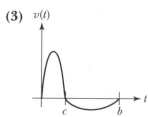

**b.** Sketch the graph of a velocity function that satisfies the given conditions.

**(1)** The associated distance function has one turning point between $t = a$ and $t = b$. The accumulated distance is positive.

(2)  The velocity is 0 at $t = a$ and at $t = b$. The accumulated distance is zero.

In previous tasks in this unit, you were interested in calculating the accumulated distance an object moves during the entire time period or the area under a curve over a fixed interval. In the next task, instead of calculating the accumulated amount over the whole interval, you will calculate a portion of the total amount or the amount-to-date. For instance, instead of calculating the accumulated distance an object moves during a 4-minute period, you might calculate the accumulated distance after 1 minute or after 2.5 minutes, or more generally, after $x$ minutes, where $x$ is any value between 0 and 4. This is not a new idea. In Task 5-13, you examined the accumulation function associated with a function whose graph lies above the horizontal axis. In the next task, you will use the same approach that you developed in Task 5-13, but this time the graph of the underlying function will be allowed to dip below the axis. You will observe how this affects the value of the accumulation function, and you will investigate how to:

- Represent the value of an accumulation function graphically by considering the region bounded by the given function.
- Approximate the value of an accumulation function using a partial Riemann sum.
- Represent the value of an accumulation function using a definite integral.

Before tackling the next task, you should review what you know about accumulation functions by reading once again the Overview for this unit and the discussion that precedes Task 5-13.

## Task 7-4: Using Partial Riemann Sums to Approximate Accumulation Functions

1.  Consider an example. Suppose an object moves at the rate of $-t + 4$ m/sec for a period of six seconds—that is, $v(t) = -t + 4$ for $0 \le t \le 6$. Let $D$ be the associated accumulated distance function. In parts a to d, calculate some values of $D$, sketch its graph, and represent $D$ by a definite integral.

**a.** Sketch the graph of $v$ for $0 \le t \le 6$.

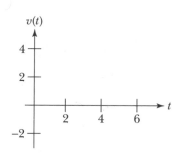

**b.** Calculate some numerical values of $D(x)$. Use a geometric approach to find the area of the appropriate region bounded by the graph of $v$, noting that the region below the axis contributes a negative amount to the value of $D(x)$. Represent each value by a definite integral.

| $x$ | 0 | 1 | 2 | 3 | 4 | 5 | 6 |
|---|---|---|---|---|---|---|---|
| $D(x)$ (numerical value) | | | | | | | |
| $D(x)$ (represented by a definite integral) | | | | | | | |

**c.** Use the information in the table to sketch the graph of the accumulation function $D(x)$ for $0 \le x \le 6$.

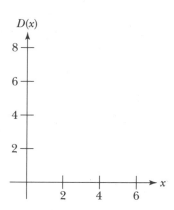

**d.** Suppose $x$ is any value between 0 and 6. Then, $D(x)$ is the accumulation function associated with the function $v$. Represent $D(x)$ by a definite integral.

$$D(x) = \int$$

**2.** Next, consider the general case. Use a partial Riemann sum to approximate the value of an accumulation function. Use a definite integral to represent the exact value.

Suppose $f$ is a continuous function defined on the closed interval $[a, b]$, where the graph of $f$ dips below the horizontal axis. Let $A$ be the accumulation function generated by $f$. Then, by definition,

$$A(x) = \int_a^x f(t)\, dt \quad \text{for } a \le x \le b.$$

**a.** Draw a squiggly curve, representing $f$ for $a \le x \le b$. Let $x$ be an arbitrary value between $a$ and $b$.

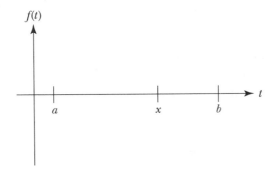

**(1)** Shade the region corresponding to $A(x)$.

**(2)** Use a different pattern to shade the region corresponding to $A(b)$.

**(3)** Find the value of $A(a)$. Justify your response.

**b.** Use the Riemann sum approach to approximate the value of $A(x)$.

**(1)** On the axes below, sketch another graph of $f$. As usual, assume $\Delta x = (b - a)/n$ and $x_i = a + i\Delta x$, for $0 \le i \le n$. Label $x_1$, $x_2$, $x_3$, and $x_n$, and draw the associated rectangles using the right endpoint of each subinterval.

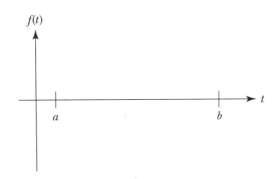

**(2)** Consider $A(x_2)$.

**(a)** Shade the region corresponding to $A(x_2)$.

**(b)** Give a formula that approximates the value of $A(x_2)$ by summing the areas of the rectangles that cover the region corresponding to $A(x_2)$. That is, give the formula for the associated *partial Riemann sum.*

**(3)** For each value of $x$ in the table given below:

**i.** In column 3, represent the exact value of $A(x)$ by a definite integral.

**ii.** In column 4, approximate $A(x)$ using a partial Riemann sum.

**iii.** In the last column, column 5, illustrate the relationship between the exact and the approximate values $A(x)$. Sketch a small graph of $f$. Outline the rectangles associated with the approximate value. Shade the region associated with the exact value.

To help you get started, a few entries in the table have been filled in for you.

| $x$ | $f(x)$ | $A(x)$ (definite integral) | Approximate value of $A(x)$ (partial Riemann Sum) | $A(x)$ (graphical rep.) |
|---|---|---|---|---|
| $x_1$ | $f(x_1)$ | | | |
| $x_2$ | $f(x_2)$ | | $f(x_1)\Delta x + f(x_2)\Delta x$ | |
| $x_3$ | $f(x_3)$ | $\int_a^{x_3} f(x)\, dx$ | | |
| ⋮ | ⋮ | ⋮ | ⋮ | ⋮ |
| $x_n$ | $f(x_n)$ | | $\sum_{i=1}^{n} f(x_i)\Delta x$ | |

Next, examine another situation where the Riemann sum approach applies. You know how to find the area of a region bounded by a curve and a closed interval. But what about the area of a region bounded by two curves, say, a "top" curve and a "bottom" curve? For instance, what if you wanted to find the area of a property bounded on the northern and southern boundaries by two meandering streams? In this case, when you try to approximate the area using a rectangular approach, the height of each rectangle is determined by the distance between the two curves, instead of, as in the case of a single curve, the distance between the horizontal axis and the curve. Consequently, you will need to think about how to find the area of a typical rectangle, where the top and bottom of the rectangle are both determined by curves.

In the next task, you will examine this situation while thinking about finding the area of a property bounded by two streams.

## Task 7-5: Representing the Area Between Two Curves by a Definite Integral

In the diagram below, suppose $f$ and $g$ represent two streams, where the older of the two streams, $g$, has more meanders. You are interested in representing the area between the two streams by a definite integral. As usual, you can do this by subdividing the region, approximating the areas of the subregions using a rectangular approach, summing the results, and then taking the limit.

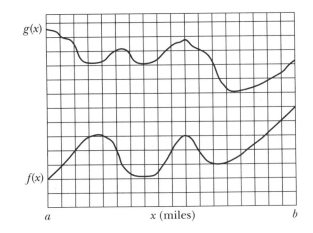

1. Partition the interval $[a, b]$ into $n$ equal subintervals with width $\Delta x$ and cover the region with rectangles as follows:

   a. On the diagram, label $a$ to be $x_0$ and label the right endpoints of the subintervals $x_1, x_2, ..., x_{i-1}, x_i, ..., x_{n-1}, x_n$. Label the width of each subinterval $\Delta x$.

   b. Let the value of $g$ at the right endpoint of a subinterval determine the *top* of the associated rectangle and the value of $f$ at the right endpoint determine the *bottom* of the rectangle. Sketch the first, $i$th, and $n$th rectangles.

2. Approximate the area between the graphs of $f$ and $g$.

   a. Find a formula for the area of the $i$th rectangle.

   b. Sum the areas of the rectangles from part a. Express your answer using sigma notation.

3. Express the exact area between the graphs in two ways:

   a. As the limit of the sum

   b. By a definite integral

4. The formula you developed in part 3 using the Riemann approach holds only if the graph of *g* lies *above* the graph of *f* for all values of *x* in the interval $[a, b]$—that is, $g(x)$ must be greater than or equal to $f(x)$, for all $a \leq x \leq b$. Explain why this must be the case.

5. If both functions are non-negative, an alternate way to find the area between two functions is to subtract the area under the bottom function from the area under the top function.

   a. On the pair of axes given below, sketch two continuous functions, *f* and *g*, where $0 \leq f(x) \leq g(x)$ for all *x* in the closed interval $[a, b]$.

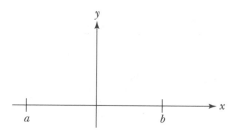

   b. Shade the region under the graph of the upper function *g* and over the closed interval $[a, b]$. Represent this area by a definite integral.

   c. Using a different pattern, shade the region under the graph of *f* and over the closed interval $[a, b]$. Represent this area by a definite integral.

   d. Represent the area between the two functions as the difference of two definite integrals.

**e.** Compare your answer to the previous question, part d, to your response to part 3b. Based on your observation, state a property of definite integrals.

**6.** Based on your work in the parts 1–5, given two functions $f$ and $g$, where $0 \leq f(x) \leq g(x)$ for all $x$ in the closed interval $[a, b]$, the area between $f$ and $g$ is represented by

$$\int_a^b (g(x) - f(x)) \, dx \quad \text{or} \quad \int_a^b g(x) \, dx - \int_a^b f(x) \, dx.$$

For these formulas to hold, the graph of $g$ must lie above the graph of $f$. However, it is not necessary for the graphs to lie above the horizontal axis. In fact, the formulas hold if the graph of $f$ or if the graphs of both $g$ and $f$ dip below the $x$-axis.

**a.** On the set of axes given below, sketch two continuous functions, $f$ and $g$, where $f(x) \leq g(x)$ for all $x$ in the closed interval $[a, b]$ and the graph of $f$ dips below the horizontal axis.

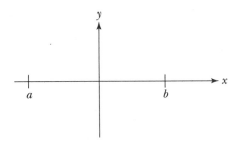

**b.** Explain why the formulas also apply in this case—that is, when the graph of $f$ lies below the horizontal axis.

*Hint: Sketch a typical rectangle determined by the two functions and find the expression for its height.*

**c.** Similarly, suppose the graphs of $f$ and $g$ both lie below the horizontal axis—that is, suppose $f(x) \leq g(x) \leq 0$ for all $x$ in the closed interval $[a, b]$. Explain why the formulas also apply in this case. Support your explanation with a sketch.

### Unit 7 Homework After Section 1

- Complete the tasks in Section 1. Be prepared to discuss them in class.

- Compare the distances traveled by two cars in HW7.1.

**HW7.1** Suppose you have two fast cars. Car 1 accelerates slowly but then goes like a rocket, whereas car 2 quickly reaches its top speed. The graphs below show the velocity of car 1, $v_1$, and the velocity of car 2, $v_2$, during the first 60 seconds.

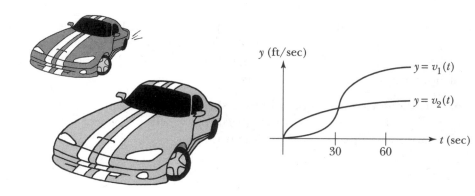

1. Using different markings, shade the regions that represent the distances car 1 and car 2 travel during the first 30 seconds. Which car travels the farthest during this time period? Justify your response.

2. Which car travels the farthest after 60 seconds? Justify your response by comparing the appropriate regions on the graph.

3. Represent the distance car 2 travels, when $15 \leq t \leq 35$, by a definite integral.

4. Suppose the two cars start at the same time, side by side, traveling in the same direction. Describe how you can use the graphs to determine the time when one car passes the other.

- Develop a formula for determining the amount of work done when the force varies continuously in HW7.2.

**HW7.2** If a constant force $F$ is applied to an object moving a distance $x$ along a straight line from point $a$ to point $b$, then the work $W$ done on the object is found by multiplying the force and the distance. That is,

$$\text{work} = \text{force} \cdot \text{distance}.$$

What if the force varies? How much work is done in this case? If you push an object from point $a$ to point $b$ and the force varies continuously, the graph of the force $F$ for $a \leq x \leq b$ might look like

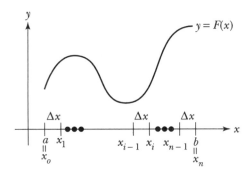

Use the Riemann sum approach to find a formula for the work done in this case. As usual, divide the distance interval $[a, b]$ into $n$ subintervals, where $\Delta x = (b - a)/n$ and $x_i = a + i\Delta x$ for $1 \le i \le n$.

1. Explain why $F(x_i)\Delta x$ approximates the amount of work done moving the object from $x_{i-1}$ to $x_i$. Shade the associated rectangle on the graph of $F$.

2. Find a formula, in terms of the limit of a sum, for the work $W$ done moving an object from $x = a$ to $x = b$ with force $F$. Carefully explain why your formula holds.

3. Express the formula from part 2 in terms of a definite integral.

• Develop a formula for calculating the mass of an object when the density varies continuously in problem HW7.3.

**HW7.3** Consider the following bar:

If you know the density of an object and its volume, then you can calculate its mass by multiplying the density and the volume. That is,

$$\text{mass} = \text{density} \cdot \text{volume}.$$

1. Suppose the density of the bar pictured above is 0.5 lb/in$^3$. Find the mass of bar.

2. Now suppose you have the same bar as pictured above, but that the density varies continuously as you move along the bar from one end to the other. In particular, suppose $D$ is the density when you are $s$ inches from the left end of the bar, where the values of $D$ at 6" intervals are

| $s$ | 0 | 6 | 12 | 18 | 24 | 30 | 36 | 42 | 48 |
|---|---|---|---|---|---|---|---|---|---|
| $D$ | 2.5 | 3.3 | 3.4 | 3.1 | 2.8 | 2.5 | 2.1 | 1.6 | 1.0 |

Use the Riemann sum approach to approximate the mass of the bar.

> *Hint: Divide the bar into eight 6" slices. For each slice, approximate the density of the slice by using the value of D at the right endpoint, and then multiply the density times the volume of the slice. Sum the results to approximate the mass.*

3. Suppose $D(s)$ where $0 \leq s \leq 48$ gives the density of the bar pictured above when you are $s$ inches from the left endpoint of the bar, where $D$ varies continuously as you move along the bar. For instance, the graph of $D$ might look like

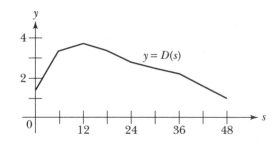

a. Use the Riemann sum approach to find a formula (in terms of the limit of a sum) for the exact mass of the bar. Carefully explain what your notation means.

b. Express the formula for the mass in terms of a definite integral.

• In HW7.4, think about different ways to represent area—namely by a verbal description, a graph, or a definite integral—and convert between the representations.

## HW7.4

1. For each of the following verbal descriptions, sketch the region, and represent its area using one or more definite integrals.

a. The region bounded by $x = -1$, $x = 1.5$, $y = x^3 + 8$, and the $x$-axis.

b. The region bounded by $f(t) = \sqrt{t}$, where $1 \leq t \leq 9$, and the $t$-axis.

c. The region bounded by the $x$-axis and
$$g(x) = \begin{cases} -x & \text{if } -2 \leq x \leq -1 \\ -x^2 + 2 & \text{if } -1 < x \leq 1. \end{cases}$$

d. The region bounded by the graphs of $h(t) = e^t$ and $g(t) = 1$, where $0 \leq t \leq 1.5$.

e. The region bounded by $y = x^3$ and $y = 4x$.

f. The region bounded by $f(x) = \sqrt{x}$, $h(x) = -\sqrt{x}$, and $x = 3$.

g. The region bounded by $w = x$, $w = x^2$, and $x = 2$.

2. Sketch the regions whose areas are given by the following definite integrals:

a. $\int_0^{2\pi} (\sin(x) + 2) \, dx$

d. $\int_{-2}^0 (-x^2) \, dx - \int_{-2}^0 (-2x) \, dx$

b. $\int_0^2 x^2 \, dx + \int_2^6 (-x + 6) \, dx$

e. $\int_{-2}^4 |x| \, dx - \int_{-2}^4 (-|x|) \, dx$

c. $\int_{-1}^2 4 \, dt - \int_{-1}^2 t^2 \, dt$

3. Represent the areas of the following shaded regions using a definite integral:

   **a.**

   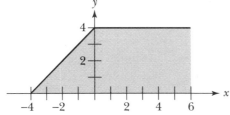

   **b.**

   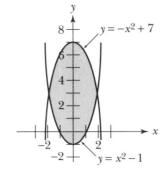

   **c.**

   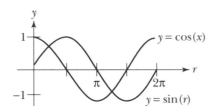

4. Consider the graphs of $y = \sin(x)$ and $y = \cos(x)$ for $0 \le x \le 2\pi$.

   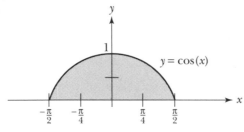

   For each of the following definite integrals:

   **i.** Make a new sketch of the graphs.

   **ii.** Shade the region associated with the integral.

   **a.** $\displaystyle\int_{0}^{\pi/4} (\cos(x) - \sin(x))\ dx$

   **d.** $\displaystyle\int_{\pi/4}^{5\pi/4} [\sin(x) - \cos(x)]\ dx$

   **b.** $\displaystyle\int_{0}^{\pi/4} \sin(x)\ dx + \int_{\pi/4}^{\pi/2} \cos(x)\ dx$

   **e.** $\displaystyle\int_{\pi}^{5\pi/4} \sin(x)\ dx + \int_{5\pi/4}^{3\pi/2} \cos(x)\ dx$

   **c.** $\displaystyle\int_{5\pi/4}^{2\pi} [\cos(x) - \sin(x)]\ dx$

• Represent some areas using definite integrals in HW7.5.

**HW7.5** You have a customer who is interested in buying a piece of property in Bellevue Park in Harrisburg, Pennsylvania. Below is a diagram showing some roads in the neighborhood, where the axes correspond to Midland and Bellevue roads. The distances are in feet.

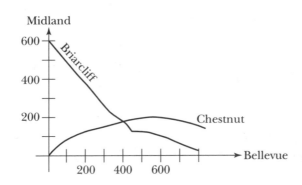

Let $b$ and $c$ be the functions corresponding to Briarcliff and Chestnut, respectively. For each of the definite integrals in parts 1 to 6, make a quick sketch of the graph and then shade the region on your sketch whose area is represented by the given integral(s).

1. $\int_0^{400} b(x)\ dx$

2. $\int_0^{400} [b(x) - c(x)]\ dx$

3. $\int_0^{400} c(x)\ dx$

4. $\int_{400}^{800} [c(x) - b(x)]\ dx$

5. $\int_{150}^{375} b(x)\ dx$

6. $\int_0^{400} b(x)\ dx - \int_0^{400} c(x)\ dx$

• Interpret some definite integrals in HW7.6.

**HW7.6**

1. Consider the graph

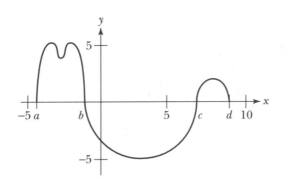

Determine whether each of the following statements is true or false. In each case, use an area-based argument to justify your conclusion.

**a.** $\int_a^b f(x)\ dx \geq \int_c^d f(x)\ dx$

**d.** $\int_a^{(a+b)/2} f(x)\ dx \approx \frac{1}{2}\int_a^b f(x)\ dx$

**b.** $\int_b^d f(x)\ dx \geq 0$

**e.** $\int_a^b f(x)\ dx \leq 25$

**c.** $\int_b^c f(x)\ dx - \int_b^0 f(x)\ dx = \int_0^c f(x)\ dx$

**2.** Consider the following graph of $g$:

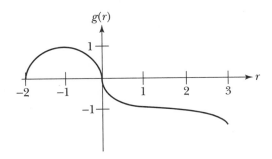

Suppose $\int_{-1}^0 g(r)\ dr = a$ and $\int_{-2}^3 g(r)\ dr = b$, where $a$ and $b$ are real numbers. Represent the value of each of the following definite integrals by an expression in terms of $a$ and $b$.

**a.** $\int_{-2}^0 g(r)\ dr$

**d.** $\int_0^3 g(r)\ dr$

**b.** $\int_0^1 g(r)\ dr$

**e.** $\int_1^3 g(r)\ dr$

**c.** $\int_{-2}^1 g(r)\ dr$

**f.** $\int_{-1/2}^1 g(r)\ dr$

- Determine the amount of heat lost and/or gained in HW7.7

**HW7.7** When the sun is shining during the day, your house collects energy through the roof. It then begins to lose energy after the sun sets. Suppose $r$ is the rate that energy is collected and lost from the roof, where $r$ is measured in ergs/hour. (*Note*: One erg is equivalent to the amount of energy released when a mosquito hits the windshield of a car that is going 60 MPH . . . at least that is what we have been told.)

Below is the graph for $r$ versus $t$ from 14 hours (2 P.M.) to 24 hours (midnight).

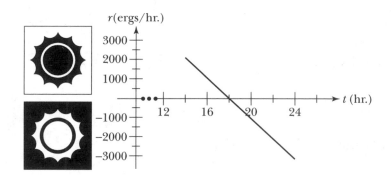

1. Find an equation for $r$ in terms of $t$.

2. When $r$ is positive, energy is being collected.

   **a.** During which time period(s) is energy being collected?

   **b.** Shade the region on the graph of $r$ that corresponds to the amount of energy collected before sunset. Explain why the region represents this amount.

   **c.** Represent the amount of energy collected by a definite integral.

   **d.** Use a geometric approach to find the amount of energy collected.

3. Similarly, when $r$ is negative, energy is being lost.

   **a.** During which time period(s) is energy being lost?

   **b.** Use a different pattern to shade the region on the graph of $r$ that corresponds to the amount of energy lost after sundown. Explain why the region represents this amount.

   **c.** Represent the amount of energy lost by a definite integral.

   **d.** Use a geometric approach to find the amount of energy lost.

4. Represent the total amount of energy collected/lost between 2 P.M. and midnight by a definite integral. Find the value of the integral.

• In HW7.8, describe the quantities represented by some values of a given accumulation function.

**HW7.8** Consider the following cumulative area function:

$$A(x) = \int_{-1}^{x} (t^2 + 1) \, dt, \quad \text{where } -1 \le x \le 4.$$

1. Sketch a graph of the function which generates $A$.

2. Using different markings, shade the following areas on the sketch you drew in part 1. Be sure and indicate which marking represents which area.

a. Shade the region whose area is equal to $A(3.5)$.

b. Shade the region whose area is given by $A(4) - A(0)$.

c. Shade the region whose area is given by $(5)(17) - A(4)$.

*Hint: Think rectangles.*

3. Determine whether each of the following statements is true or false by considering the sketch you drew in part 1. Justify your response.

a. $A(0) \geq 2$

b. $A(1) = (2)A(0)$

c. $A(4) - A(0) < \frac{1}{2}(1 + 17)(4)$

*Hint: Think trapezoids.*

4. Approximate the value of $A$ at the following points by finding the *numerical* value of the associated partial Riemann sum, using the right endpoint of the subintervals.

a. Approximate $A(1)$, with $\Delta x = 0.5$.

b. Approximate $A(0)$, with $\Delta x = 0.25$.

• As you know, the value of a definite integral may be 0. With this in mind, think some more about the meaning of "cumulative distance" in HW7.9.

**HW7.9** Assume the graph of the velocity function for a cart that moves up an inclined ramp and then back down the ramp has the following shape:

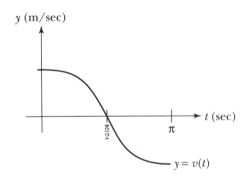

1. Assume "distance" means the distance of the cart from the bottom of the ramp. Give the time period (in terms of $\pi$) when the cart is:

a. Moving up the ramp. Justify your response.

b. Moving down the ramp. Justify your response.

2. Suppose $x$ is any time between $t = 0$ and $t = \pi$. Define the cumulative distance function $D(x)$ associated with $v$ in terms of a definite integral.

**3.** For each situation described below in parts a through c, represent the cumulative distance that the cart travels:

   **i.** in terms of $D$

   **ii.** in terms of a definite integral

   **iii.** by shading the appropriate region on the graph of $v$

   **a.** The cumulative distance from the time the cart begins to move until it turns back.

   **b.** The cumulative distance from the time the cart turns back until it stops.

   **c.** The cumulative distance for $\pi/4 \leq t \leq 3\pi/4$.

**4.** Represent the approximate value of $D(\pi/2)$ by an equation using a partial Riemann sum. Assume $\Delta t = \pi/8$.

*Note: You can not find the numerical value since you do not have an equation for v(t).*

**5.** Explain why $D(\pi) = 0$.

**6.** Sketch the distance versus time graph for the cart assuming the cart is at the bottom of the ramp at $t = 0$ seconds and the maximum distance the cart travels from the starting point is 1.5 meters.

---

## SECTION 2

### Calculating Antiderivatives

In Units 5 and 6, you thought about what the derivative of a function represents and you learned how to calculate derivatives. You developed methods for finding derivatives of algebraic expressions involving polynomials, and trigonometric, exponential, and logarithmic functions. You became comfortable using the rules, including the Extended Power Rule, Quotient Rule, Product Rule, and Chain Rule.

    The goal of this section is to explore ways to reverse the differentiation process. You can think of it as being given the answer to a differentiation problem and then being asked to go up a step to find the original function. This process is called *antidifferentiation*, where the prefix "anti" means before, opposite, or reverse. For example, consider

$$f(x) = 4x^3.$$

Suppose you are asked to find a function *whose derivative* is $4x^3$. One possibility is

$$F(x) = x^4$$

since

$$F'(x) = f(x) \quad \text{or} \quad \frac{d}{dx}(x^4) = 4x^3.$$

In this case, $F$ is said to be an *antiderivative* of $f$.

There is one slight hitch to reversing the process and finding an anti-derivative. When you are asked to find the derivative of a function, there is only one possible answer. For example, the derivative of $F(t) = t^2 + 6$ is the function $2t$. That is it. There are no other possibilities. However, when you are asked to antidifferentiate a function, there are many right answers. For example, when asked to antidifferentiate $f(x) = 4x^3$, then $x^4$ is a correct re-sponse, but so are

$$x^4 + 89, \quad x^4 - 100.9, \quad \text{and} \quad 2\pi + x^4$$

since the derivative of each of these functions is $4x^3$. Consequently, in ad-dition to finding antiderivatives, you need to think about how all the anti-derivatives of a particular function relate to one another. How are their ex-pressions related? How are their graphs related? This is the goal of the next task.

Note that, in general, lowercase letters will refer to a given function and upper case letters will denote antiderivatives of a function.

## Task 7-6: Examining How Antiderivatives Are Related

1. A differentiable function has one and only one derivative. A function, however, has many antiderivatives. Examine how the expressions repre-senting the antiderivatives of a function are related.

   a. Consider the functions $F_1$ through $F_6$, where

   $$F_1(t) = t^2, \qquad F_2(t) = t^2 + 3,$$

   $$F_3(t) = t^2 + 5.6, \qquad F_4(t) = t^2 - 2.8,$$

   $$F_5(t) = t^2 - 3, \qquad F_6(t) = t^2 + C, \text{ where } C \text{ is a real number}$$

   (1) $F_1$ through $F_6$ are all antiderivatives of $f(t) = 2t$. Explain why this is the case.

   (2) List three other antiderivatives of $f(t) = 2t$.

   (3) Describe how the expressions representing the various anti-derivatives of $f(t) = 2t$ are related.

**b.** Consider $g(x) = (x + 3)^5$.

  **(1)** Show that $G(x) = \frac{1}{6}(x + 3)^6 + 7$ is an antiderivative of $g$.

  **(2)** List three other antiderivatives of $g$.

**c.** Generalize your observations. Describe how the expressions representing the antiderivatives of a given function are related.

**2.** Next, examine how the graphs representing the antiderivatives of a function are related.

**a.** Consider the six antiderivatives of $f(t) = 2t$ listed in part 1a, namely $F_1$ through $F_6$.

  **(1)** On the axes given below, roughly sketch the graphs of $F_1$ through $F_6$.

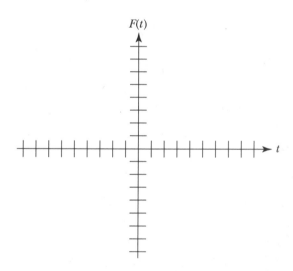

  **(2)** Describe how the graphs of the antiderivatives of $f(t) = 2t$, which you sketched above, are related.

**b.** Suppose $F$ is the antiderivative of some function $f$, where the graph of $F$ is given below. Sketch the graphs of two other antiderivatives of $f$.

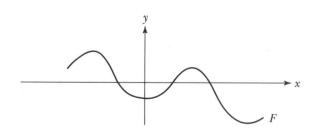

**c.** Generalize your observations. Describe how the graphs of the antiderivatives of a given function are related.

Obviously, you cannot write down all the antiderivatives of a given function, since every real number $C$ gives you a different one. You indicate that this is the case by tacking a $C$ on to the basic solution. For instance, you can represent the set of all antiderivatives of $f(t) = 2t$ by

$$F(t) = t^2 + C, \text{ where } C \text{ is any real number.}$$

In this case, $F$ is said to be the *general antiderivative* of $f$.

The antiderivative of a function is another function, just as the derivative of a function is a function. If $F$ is the general antiderivative of $f$, then $F$ can also be represented using the notation for an *indefinite integral*, namely

$$\int f(x) \ dx = F(x).$$

For instance, if $f(t) = 2t$, then the general antiderivative of $f$ is $t^2 + C$ and you can write

$$\int 2t \ dt = t^2 + C,$$

which is read, "The integral—or the antiderivative—of $2t$ equals $t^2 + C$."

The good news is that there is an easy way to check if your solution is right: simply differentiate your result to make sure it gives you what you want. For instance,

$$\frac{d}{dt}(t^2 + C) = 2t,$$

so you know that $t^2 + C$ is the (general) antiderivative of $2t$.

In the next task, you will develop rules for finding the antiderivatives of power, trigonometric, and exponential functions. Before you begin, review the rules for finding the derivative of $k$, where $k$ is a constant, $x^n$, $e^x$, $\ln(x)$, $\sin(x)$, and $\cos(x)$.

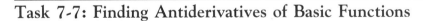

## Task 7-7: Finding Antiderivatives of Basic Functions

**1.** The indefinite integral notation should look familiar.

**a.** Explain how the notation for an indefinite integral differs from the notation for a definite integral.

**b.** Why do you think one type of integral is called "definite," and the other is called "indefinite"?

2. Discover some useful rules by finding the antiderivatives of the functions given in parts a through o below. Use the following process:

**Step 1:** Ask yourself, "What function would I differentiate to get the given function?"

For example, if you want to find the antiderivative of $x^2$—that is, if you want to evaluate

$$\int x^2 \, dx$$

—ask yourself, "What function would I differentiate to get $x^2$?"

After exploring various possibilities, you should conclude that

$$\int x^2 \, dx = \frac{x^3}{3} + C.$$

*Note:*

- Watch the names of the variables. Your answer should be written in terms of the same variable as the given function.
- Find the general antiderivative. (Do not forget "+C.")

**Step 2:** Check that your answer is correct by differentiating it and showing that you get back to where you started.

For example, $\frac{x^3}{3} + C$ is the general antiderivative of $x^2$, since

$$\frac{d}{dx}\left(\frac{x^3}{3} + C\right) = x^2.$$

Use these two steps to find the antiderivatives of the following functions:

**a.** $\int 6 \, dt$

Check:

**b.** $\int -10.89 \, dr$

Check:

**c.** $\int dx$

Check:

**d.** $\int 2x\, dx$

Check:

**e.** $\int x\, dx$

Check:

**f.** $\int 3x^2\, dx$

Check:

**g.** $\int -3t^{-4}\, dt$

Check:

**h.** $\int t^{-8}\, dt$

Check:

**i.** $\int \dfrac{1}{x}\, dx$

Check:

**j.** $\int \dfrac{-4.5}{x}\, dx$

Check:

**k.** $\int e^x\, dx$

Check:

**l.** $\int -2e^x\, dx$

Check:

**m.** $\int -\sin(x)\, dx$

Check:

**n.** $\int 3\sin(r)\, dr$

Check:

**o.** $\int \cos(t)\, dt$

Check:

3. Generalize your observations from part 2 by finding a formula for each of the following indefinite integrals. Check that your formula is correct by differentiating it.

**a.** $\int k \, dx$, where $k$ is a constant

Check:

**b.** $\int x^n \, dx$, when $n \neq -1$

Check:

**c.** $\int \frac{1}{x} \, dx$

Check:

**d.** $\int e^x \, dx$

Check:

**e.** $\int \sin(x) \, dx$

Check:

**f.** $\int \cos(x) \, dx$

Check:

**4.** Summarize the rules for finding derivatives and antiderivatives of basic functions. For each basic function in row 2, find its general antiderivative and its derivative and enter the results in the column. To help you get started, we have completed the column for $f(x) = x^n$, $n \neq -1$. Assume that $k$ is any real number.

Differentiation / Integration

| Antiderivative | $\int f(x)\,dx$ | | $\dfrac{x^{n+1}}{n+1} + C$ | | | | |
|---|---|---|---|---|---|---|---|
| Basic Function | $f(x)$ | $k$ | $x^n,\ n \neq -1$ | $\dfrac{1}{x}$ | $e^x$ | $\sin(x)$ | $\cos(x)$ |
| Derivative | $f'(x)$ | | $nx^{n-1}$ | | | | |

5. Use your understanding of the relationship between integration and differentiation to explain why each of the following statements is true:

   a. $\dfrac{d}{dx}\left(\int f(x)\,dx\right) = f(x)$

   b. $\int f'(x)\,dx = f(x) + C$

At this point, you can find antiderivatives of basic functions, such as $x^3$ and $\sin(x)$, but what about finding the antiderivative of a combination of functions, such as $\int (3\sin(r) - 2\cos(r) + r^2/4)\,dr$?

   In the next task, you will investigate how to find the antiderivatives of a scalar multiple of a function and the sum of two or more functions. Since integration and differentiation are inverse processes, it should not be surprising that finding the antiderivatives of these types of combinations is directly related to finding their derivatives, namely:

• The derivative of constant times a function equals the constant times the derivative of the function.

• The derivative of a sum equals the sum of the derivatives.

## Task 7-8: Finding Antiderivatives of Linear Combinations

**1.** The Scalar Multiple Rule for Antiderivatives states that

---

**Scalar Multiple Rule for Antiderivatives**

Let $f$ be a function and $k$ be a fixed constant.

- $\int kf(x)\ dx = k \int f(x)\ dx$

- "The integral of a constant times a function equals the constant times the integral of the function."

---

This means that you can "pull" a constant outside an integral. For example, to find $\int 4x^3\ dx$, you can first find $\int x^3\ dx$ and then multiply the result by 4, since $\int 4x^3\ dx = 4 \int x^3\ dx$.

To show that the claim holds, it suffices to show that the derivatives of both sides are equal, that is,

$$\frac{d}{dx}(\int kf(x)\ dx) = \frac{d}{dx}(k \int f(x)\ dx).$$

This can be done by showing that the derivatives are equal to the same thing, which in this case is $k\,f(x)$.

In parts a and b below, do this, using the facts that

**i.** $\dfrac{d}{dx}$ and $\int$ are inverse operations.

**ii.** The derivative of a constant times a function equals the constant times the derivative of the function.

**a.** Explain why $\dfrac{d}{dx}(\int kf(x)\ dx) = k\,f(x)$.

**b.** Explain why $\dfrac{d}{dx}(k \int f(x)\ dx) = k\,f(x)$.

**2.** The Sum Rule for Antiderivatives states that

---

**Sum Rule for Antiderivatives**

Let $f$ and $g$ be given functions.

- $\int (f(x) + g(x))\ dx = \int f(x)\ dx + \int g(x)\ dx$

- "The integral of a sum equals the sum of the integrals."

---

For example, to find $\int (x^2 + 6x)\, dx$, you can first find $\int x^2\, dx$ and $\int 6x\, dx$ and then add the results, since

$$\int (x^2 + 6x)\, dx = \int x^2\, dx + \int 6x\, dx.$$

As with the Scalar Multiple Rule, which you considered in part 1, to show that this claim holds, it suffices to show that the derivatives of both sides are equal, that is,

$$\frac{d}{dx}\left(\int (f(x) + g(x))\, dx\right) = \frac{d}{dx}\left(\int f(x)\, dx + \int g(x)\, dx\right).$$

This can be done by showing that the derivatives are equal to the same thing, which in this case is $f(x) + g(x)$.

In parts a and b below, do this. As before, use the fact that $\frac{d}{dx}$ and $\int$ are inverse operations. In addition, use the fact that the derivative of the sum of two functions equals the sum of the derivatives of the functions.

a. Explain why $\dfrac{d}{dx}\left(\int (f(x) + g(x))\, dx\right) = f(x) + g(x)$.

b. Explain why $\dfrac{d}{dx}\left(\int f(x)\, dx + \int g(x)\, dx\right) = f(x) + g(x)$.

3. Evaluate the following indefinite integrals.

   i. First, think of the given integral as a sum, where each term in the sum consists of a constant (which may be 1 or a negative number) times the integral of one of the basic functions that you considered in the last task.

   ii. Next, use the rules for the basic functions to find the antiderivative of each term.

   iii. As always, check your answers by differentiating.

   a. $\int (3x^2 + x)\, dx$

   Check:

**b.** $\int (-3t^{-4} + t^{-8} - \frac{4}{t}) \, dt$

Check:

**c.** $\int (3 \sin(r) - 2.5 \cos(r) + \frac{2}{3}e^r) \, dr$

Check:

**4.** When you evaluate the integral of a sum by first expressing the integral as the sum of a collection of integrals, theoretically each term in the sum contributes a "$+C$" to the final answer. Explain why it is okay to combine all the "$+Cs$" into one constant term.

At this point you can find the antiderivative of a linear combination of functions, using the rules in the table given below. In Unit 8, you will develop techniques for finding antiderivatives of some other combinations.

|  | | | |
|---|---|---|---|
| Differentiation Integration | Antiderivative | $k \int f(x) \, dx$ | $\int f(x) \, dx + \int g(x) \, dx$ |
| | Combination | $kf(x)$ | $f(x) + g(x)$ |
| | Derivative | $kf'(x)$ | $f'(x) + g'(x)$ |

The "+ C" portion of an antiderivative is a way of indicating that a given function has an antiderivative corresponding to each real number. Sometimes, however, you want to find a specific antiderivative, whose graph passes through a given point. This is what the next task is all about.

## Task 7-9: Finding Specific Antiderivatives

1. Suppose $M$ is an antiderivative of $f(t) = 2t$ where the graph of $M$ passes through the point $P(1, 12.5)$.

   a. Represent $M$ by an expression.

   *Hint: First find the general antiderivative of f. Then, find a value for the constant C, so that M(1) = 12.5.*

   b. Is your $M$ unique? That is, is your answer to part a the only function that is an antiderivative of $f(t) = 2t$ which passes through the point $P(1, 12.5)$? Justify your response.

2. Suppose $G(x) = \int \dfrac{-4}{x}\, dx$, where $G(1) = 3$. Find $G$.

3. Recall that the values of any two antiderivatives of a given function differ by a constant (see Task 7-6).

   a. Find two antiderivatives, $H_1$ and $H_2$, of $h(x) = x^2 + 2x - 1$, so that $H_2(x) - H_1(x) = 6$ for all real numbers $x$.

**b.** Consider $g(x) = -3e^x + 1$. Suppose $A$ is the antiderivative of $g$ such that $A(0) = 0$ and $F$ is the antiderivative of $g$ such that $F(1) = -3e$.

**(1)** Represent $A$ by an expression.

**(2)** Represent $F$ by an expression.

**(3)** Find the value of $K$ so that $F(x) = A(x) + K$ for all $x$.

**4.** Recall that the vertical distance between the graphs of any two anti-derivatives of a given function is constant (see Task 7-6).

   **a.** Suppose $Q$ is the antiderivative of $q$, where the graph of $Q$ is given below. Suppose $P$ is another antiderivative of $q$, where $P(2) = -1.5$. Sketch the graph of $P$.

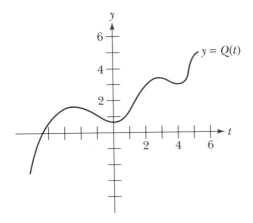

   **b.** On the axes given below, sketch the graph of a function that satisfies the specified conditions. Label each of the graphs.

     **(1)** Suppose $h$ is a function where $h(0) = 10$ and $h' = f'$. Thus, $h$ and $f$ are two antiderivatives of the same function. Sketch the graph of $h$.

     **(2)** Suppose $g$ is a function where $g(2) = -10$ and $g' = f'$. Sketch the graph of $g$.

**(3)** Suppose $k$ is a function where $k(-2) = f(-2)$ and $k' = f'$. Sketch the graph of $k$.

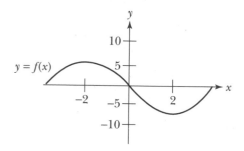

5. Consider the following graph of $y = v(t)$:

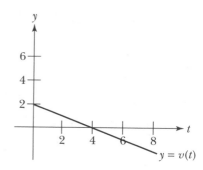

**a.** Suppose $s$ is the antiderivative of $v$, where $s(4) = 6$. Find $s$.

**b.** Since $s$ is an antiderivative of $v$, $v$ is the derivative of $s$. Consequently, you can use $v$ to find the intervals where $s$ is increasing and decreasing, and you can use $v$ to find the $t$-values of the relative extrema of $s$. Use $v$ to describe the shape of the graph of $s$.

**c.** Sketch the graph of $s$ on the same pair of axes as $v$.

## Unit 7 Homework After Section 2

- Complete the tasks in Section 2. Be prepared to discuss them in class.

- Evaluate some antiderivatives in HW7.10.

**HW7.10** Evaluate the following antiderivatives using the rules developed in this section. You may need to do some algebraic manipulations beforehand. Differentiate your answer to show that it is correct.

1. $\int -4e^x \, dx$

2. $\int (5n - 3)^2 \, dn$

3. $\int \dfrac{dx}{x^3}$ or $\int \dfrac{1}{x^3} \, dx$

4. $\int \dfrac{x^3 - 4x^2 + 9x}{x} \, dx$

5. $\int \dfrac{-2.5}{t} \, dt$

6. $\int \dfrac{r^2 - 4}{r + 2} \, dr$

7. $\int (r^{1/4} - r^4) \, dr$

8. $\int (t^3 - 2t)(2t + 3) \, dt$

9. $\int (\sin(x) - \cos(x)) \, dx$

10. $\int (\sin^2 x - \cos^2 x) \, dx$

*Hint: First use a trig identity to simplify the expression.*

- Use antiderivatives to solve some problems in HW7.11.

**HW7.11** Find the specified functions.

1. Suppose $f'(x) = \sqrt{x} + \dfrac{1}{\sqrt{x}}$, where $f(16) = 0$. Find $f$.

2. Suppose $\dfrac{dy}{dx} = 4 \cos(x)$, where the graph of $y$ passes through the point $P(0, 10)$. Find $y$.

3. Suppose the velocity function is given by $v(t) = 2t^3$. Find the distance function $s$, where $s(1) = 3$.

4. In your French class, you discover that your rate of memorizing French words is $M'(t) = 0.2t - 0.003t^2$, where $M(t)$ is the number of words you are able to memorize in $t$ minutes.

   **a.** Find $M(t)$, where $M(0) = 0$.

   **b.** Approximately how many words can you memorize after 8 minutes? After 10 minutes?

- Compare the graph of a function and one of its antiderivatives in HW7.12.

**HW7.12** For each of the following functions:

   **i.** Find the antiderivative of $f$ where $C = 0$. Call the antiderivative $F$.
   **ii.** Fill in the table by evaluating $F$ at the indicated values.
   **iii.** Use your CAS to plot $F$ and $f$ on the same pair of axes over the indicated domain. Label the graphs.
   **iv.** Describe the relationship between the graphs of $F$ and $f$. For example, $F$ is increasing when $f$ is positive, and so on.

**1.** $f(x) = 5,\ 1 \le x \le 5$

| x | 1 | 2 | 3 | 4 | 5 |
|---|---|---|---|---|---|
| F(x) | | | | | |

**2.** $f(x) = 2x,\ -3 \le x \le 3$

| x | −3 | −2 | −1 | 0 | 1. | 2 | 3 |
|---|----|----|----|---|----|---|---|
| F(x) | | | | | | | |

**3.** $f(x) = \sin(x),\ 0 \le x \le 2\pi$

| x | 0 | $\dfrac{\pi}{4}$ | $\dfrac{\pi}{2}$ | $\dfrac{3\pi}{4}$ | $\pi$ | $\dfrac{5\pi}{4}$ | $\dfrac{3\pi}{2}$ | $\dfrac{7\pi}{4}$ | $2\pi$ |
|---|---|---|---|---|---|---|---|---|---|
| F(x) | | | | | | | | | |

- Find the antiderivative of the general exponential function in HW7.13.

**HW7.13**

**1.** Show that $\dfrac{d}{dx}\left(\dfrac{1}{\ln a}a^x\right) = a^x$, where $a$ is a positive real number.

*Hint: Note that ln a is constant, and recall that $\dfrac{d}{dx}\,(a^x) = (\ln a)a^x$ (see Task 6-16).*

**2.** Consequently, since integration reverses the process of differentiation,

$$\int a^x\,dx = \frac{1}{\ln a}a^x + C.$$

Use this new integration rule to find the following antiderivatives. As usual, you may have to do some algebraic manipulation before applying the rule.

**a.** $\displaystyle\int 3^x\,dx$

**b.** $\displaystyle\int (10)(10^w)\,dw$

**c.** $\displaystyle\int \frac{4^p}{\ln 4}\,dp$

**d.** $\displaystyle\int (5x + 5^x)\,dx$

**e.** $\displaystyle\int 2\left(\frac{1}{2}\right)^m dm$

**f.** $\displaystyle\int 6^{t+1}\,dt$

**3.** Find the area of the region under the graph of $y = 5^x$ from $x = 0$ to $x = 2$. Sketch a graph of the function and shade the region.

**4.** Find the area of the region bounded by the graphs of $y = 2^x$, $y = 1 - x$, and $x = 1$. Sketch a graph of the functions and shade the region.

• Consider the integration rules corresponding to the rules for differentiating $\sec(x)$, $\csc(x)$, $\tan(x)$, and $\cot(x)$ in HW7.14.

**HW7.14** Recall that

$$\frac{d}{dx}(\csc(x)) = -\csc(x)\cot(x), \qquad \frac{d}{dx}(\sec(x)) = \sec(x)\tan(x),$$

$$\frac{d}{dx}(\tan(x)) = \sec^2(x), \qquad \frac{d}{dx}(\cot(x)) = -\csc^2(x).$$

(See HW6.16.) Consequently,

$$\int \csc(x)\,\cot(x)\,dx = -\csc(x) + C, \qquad \int \sec(x)\,\tan(x)\,dx = \sec(x) + C,$$

$$\int \sec^2(x)\,dx = \tan(x) + C, \qquad \int \csc^2(x)\,dx = -\cot(x) + C.$$

Use these new integration rules to find the following antiderivatives. You may have to do some algebraic manipulation before applying a rule.

**1.** $\displaystyle \int (4\sec^2(x) + \csc(x)\,\cot(x))\,dx$

**2.** $\displaystyle \int \sec(r)\,(8\sec(r) + 5\tan(r))\,dr$

**3.** $\displaystyle \int (\tfrac{1}{2}\cot(\theta) - \csc(\theta))\csc(\theta)\,d\theta$

**4.** $\displaystyle \int (\sqrt{x} - \csc^2(x))\,dx$

**5.** $\displaystyle \int 2\left(\sec^2(x) + \frac{1}{x^2}\right)dx$

**6.** $\displaystyle \int \frac{(\sqrt{t}\sec(t))^2}{3t}\,dt$

• Use your CAS to find some antiderivatives in HW7.15.

**HW7.15** Figure out how to use your CAS to find the antiderivative of a function. Then, for each of the following functions (most of which you do not know how to solve by hand, at least not yet):

   **i.** Find the antiderivative of the function (where $C = 0$) using your CAS.

   **ii.** Check your CAS answer by differentiating it using pencil and paper.

**1.** $f(x) = 9x^{1/2} - 25x^{-6} + 32$

**4.** $f(r) = \dfrac{2r}{r^2 - 1}$

**2.** $g(x) = \cos(x)\sin^{10}(x)$

**5.** $h(r) = 2r\sqrt{r^2 - 9}$

**3.** $v(t) = -20e^t + 2.5\sin(t)$

**6.** $k(t) = -4t^3\sin(t^4)$

## SECTION 3

## Fundamental Theorem of Calculus

In Unit 5, you developed the following definition of the definite integral of a function $f$ defined on the interval $[a, b]$:

$$\int_a^b f(x)\ dx = \lim_{n \to \infty} \sum_{i=1}^n f(x_i)\Delta x,$$

where $\Delta x = (b - a)/n$ and $x_i = a + i\Delta x$, $0 \le i \le n$. You explored ways to approximate the value of the definite integral using the Riemann sum approach and were able to make your approximations more and more precise by increasing the number of subdivisions of $[a, b]$. At this point, you know what a definite integral is, you are acquainted with situations where definite integrals are useful, and you know how to approximate them. But, you do not know a direct way to evaluate them.

One of the primary goals of this unit is to find a way to evaluate definite integrals that does not involve evaluating the limit of a sum. In preparation for this, you have looked carefully at two functions associated with a given function $f$, namely:

**1.** The accumulation function $A$ generated by $f$:

$$A(x) = \int_a^x f(t)\ dt \quad \text{for } a \le x \le b$$

**2.** The antiderivative $F$ of $f$:

$$F(x) = \int f(x)\ dx, \quad \text{where } F'(x) = f(x)$$

These two functions, the accumulation function and the antiderivative function, are related in a powerful way. If $A$ is the accumulation function generated by $f$, then

*A is an antiderivative of f.*

Another way of saying this is

The derivative of $A$ equals $f$ or $A'(x) = f(x)$.

This important fact is part of the *Fundamental Theorem of Calculus (FTC)*, and it provides a direct way to evaluate definite integrals.

In Task 5-13 and again in HW5.13, you showed that $A'(x) = f(x)$, for some linear functions. How did you do this? First, you found some values of $A$ by using a geometric approach to calculate the areas of the corresponding regions bounded by the graph of $f$. You used these values to sketch the graph of $A$ and to find an expression for $A(x)$. Next, you used the definition of derivative (this was before you learned the rules for differentiation) to find the derivative of $A$ and in the process showed that $A'(x) = f(x)$, which was what you wanted to do.

This strategy worked since $f$ was linear and you were able to find an expression for $A$ by looking at its graph. But, what if $f$ is more complicated? How might you convince yourself that the accumulation function $A$ is an antiderivative of $f$? One approach is to compare the graph of $A$ to the graph of an antiderivative of $f$. If the FTC is true, then

- The graph of $A$ and the graph of an antiderivative will have similar shapes and the vertical distance between the two graphs will be constant.
- The value of $A$ and the value of the antiderivative at each point will differ by a constant.

In the next task, you will use your CAS and this approach to show that the FTC appears to hold for several nonlinear functions.

---

## Task 7-10: Using Your CAS to Compare Accumulation Functions and Antiderivatives

1. Determine how to use your CAS to represent the accumulation function $A$ generated by a given function $f$, where

$$A(x) = \int_a^x f(t) \, dt \quad \text{for } a \le x \le b.$$

   Note that $A$ is defined as a function of $x$ and the variable $t$ is the (dummy) variable of integration.

   For example, in *Mathematica*, if you first define $a$ and $f$, you can define $A$ as follows:

   ```
   a := ...
   f[x_] := ...
   A[x_] := Integrate [f[t], {t, a, x}]
   ```

   Since $A$ is a function, you can graph and evaluate $A$ in the usual way. Observe how closely the *Mathematica* syntax mimics the mathematical definition of $A$.

2. Consider $f(x) = 2x$ for $1 \le x \le 3$. Compare the accumulation function generated by $f$ to an antiderivative of $f$.
   **a.** Let $F$ be the antiderivative of $f$, where $C = 0$.

      **(1)** Find an expression for $F$.

      **(2)** Define $F$ using your CAS.

   **b.** Let $A$ be the accumulation function generated by $f$. Define $A$ using your CAS.

   **c.** Show that the values of the accumulation function $A$ and the antiderivative $F$ differ by a constant.

(1) Fill in the table below, using your CAS to evaluate $A$ at the indicated values.

| x | 1 | 1.5 | 2 | 2.5 | 3 |
|---|---|-----|---|-----|---|
| A(x) | | | | | |
| F(x) | | | | | |

(2) Compare the values of $A$ and $F$. If $A$ is also an antiderivative $f$, then the values of $A(x)$ and $F(x)$ should differ by a constant for any value of $x$. If this is not the case, stop and figure out what is wrong.

(3) Find the value of the constant $K$ so that $F(x) = A(x) + K$ for all $x$ between 1 and 3.

d. Show that the graphs of the accumulation function $A$ and the antiderivative $F$ have similar shapes.

(1) Plot the functions $A$ and $F$ on the same pair of axes, for $1 \leq x \leq 3$. Print a copy of the graphs and place it below. Label the graphs.

(2) If $A$ is also an antiderivative $f$, then the vertical distance between the graphs of $A$ and $F$ should be constant for any value of $x$. If this is not the case, stop and figure out what is wrong.

3. Examine the relationship between the accumulation function and an antiderivative for the nonlinear functions listed below. For each function in parts a through c:

i. Let $F$ be the antiderivative of $f$, where $C = 0$. Find an expression for $F$. Define $F$ using your CAS.

ii. Let $A$ be the accumulation function generated by $f$. Define $A$ using your CAS.

iii. Show that the values $A$ and $F$ differ by a constant.

(1) Calculate the values of $A$ and $F$ at four or five different points in the given interval, including the endpoints of the interval. Enter the values in a table.

(2) Find the value of the constant $K$ so that $F(x) = A(x) + K$ for all $x$.

**iv.** Show that the graphs of $A$ and $F$ have similar shapes. Plot $A$ and $F$ on the same pair of axes, over the given interval. Print a copy of the graphs and place it in your activity guide. Label the graphs.

**a.** $f(x) = 3x^2 + 6,\ -2 \le x \le 3$

**b.** $f(x) = \cos(x),\ 0 \le x \le 2\pi$

**c.** $f(x) = e^x,\ -1 \le x \le 3$

Based on your observations in the last task, the claim that the accumulation function $A$ generated by a function $f$ is an antiderivative of $f$ should seem reasonable. Unfortunately, showing that a claim holds for a few examples does not imply that it is always true. You need to show that the relationship holds when $f$ is *any* continuous function and $A$ is the associated accumulation function. One way to do this is to show that the derivative of $A$ equals $f$, or $A'(x) = f(x)$. This is the approach you have been using to check that the answer which you get when you evaluate an antiderivative is correct. In the case of each of the antiderivatives you have evaluated to date, you were able to use the rules of differentiation to check the answer. However, in the case of the accumulation function none of the rules applies—that is, you do not know a rule for finding $A'(x)$ when

$$A(x) = \int_a^x f(t)\ dt \quad \text{for } a \le x \le b.$$

However, you can use the definition of derivative to find $A'(x)$, where

$$A'(x) \stackrel{\text{def.}}{=} \lim_{h \to 0} \frac{A(x+h) - A(x)}{h}.$$

Consequently, to show that $A'(x) = f(x)$, you need to show that

$$A'(x) \stackrel{\text{def.}}{=} \lim_{h \to 0} \frac{A(x+h) - A(x)}{h} \stackrel{?}{=} f(x).$$

One way to convince yourself that this is true is to approximate $A(x+h) - A(x)$. For instance, consider the following diagram of an arbitrary continuous function $f$ defined on the interval $[a, b]$:

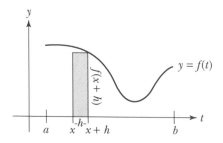

Observe that the area of the shaded rectangle equals $f(x+h) \cdot h$, and that the area of the subregion under the graph of $f$, between $x$ and $x+h$, equals $A(x+h) - A(x)$, since $A$ is the accumulation function generated by $f$. When $h$ is small, the area of the rectangle approximates the area of the subregion. That is,

$$A(x+h) - A(x) \approx f(x+h) \cdot h \quad \text{for } h \text{ small.}$$

Using this observation, we see that

$$A'(x) = \lim_{h \to 0} \frac{A(x+h) - A(x)}{h} \qquad \text{Use the definition of derivative.}$$

$$= \lim_{h \to 0} \frac{f(x+h) \cdot h}{h} \qquad \text{Replace } A(x+h) - A(x) \text{ by } f(x+h) \cdot h, \text{ since } h \text{ is small.}$$

$$= \lim_{h \to 0} f(x+h) \qquad \text{Cancel } h, \text{ since } h \ne 0.$$

$$A'(x) = f(x) \qquad \text{Take limit as } h \to 0.$$

This is exactly what we wanted to show. So, it is reasonable to believe that the accumulation function generated by a function is an antiderivative of the function. This is Part I of the Fundamental Theorem of Calculus. Part II of the FTC—which we will prove shortly, using the first part—provides you with a direct way to evaluate definite integrals. In particular, Part II claims that if $F$ is any antiderivative of $f$, then

$$\int_a^b f(t)\ dt = F(t)\Big|_{t=a}^{t=b} = F(b) - F(a).$$

Therefore, to evaluate $\int_a^b f(t)\ dt$:

- Find *any* antiderivative of $f$, say $F$.

- Evaluate $F$ at $a$ and $F$ at $b$, and subtract the results.

For example, to evaluate $\int_1^4 (t-1)\ dt$, find any antiderivative of $f(t) = t - 1$, say $F(t) = \frac{1}{2}t^2 - t$. Then, according to Part II,

$$\int_1^4 (t-1)\ dt = (\tfrac{1}{2}t^2 - t)\Big|_{t=1}^{t=4} = (\tfrac{1}{2}(4^2) - 4) - (\tfrac{1}{2}(1^2) - 1) = \frac{9}{2},$$

which is the correct answer since the area of the region bounded by $f(t) = t - 1$ over the interval $[1, 4]$ is $\frac{9}{2}$, as shown in the graph below.

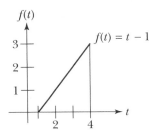

For this example, you did not need the FTC, since you could use a geometric approach. This is not always the case. For instance, you cannot find use a geometric approach to find the value of $\int_0^4 x^2\ dx$, since there is no formula for calculating the area of associated region. However, you can use Part II, since you know that $\frac{1}{3}x^3$ is an antiderivative of $x^2$.

In the next task, you will convince yourself that Part II of the Fundamental Theorem of Calculus holds by applying the theorem to some definite integrals that you already know how to evaluate and then comparing the results. After completing the task, we will look at the proof of part II and then you will apply it to some situations that cannot be solved using a geometric approach.

## Task 7-11: Testing Part II of the FTC

Show that Part II of the FTC holds for the definite integrals given in parts 1 to 3 given below. For each integral,

  a. Use a geometric approach to evaluate the definite integral.

   i. Sketch the graph of $f$, for $a \le t \le b$.

   ii. Shade the region associated with the definite integral.

   iii. Find the value of the integral, keeping in mind that a region below the horizontal axis contributes a negative amount to the value of the integral.

  b. Use Part II of the Fundamental Theorem of Calculus to evaluate the definite integral.

   i. Find an antiderivative $F$ of $f$.

   ii. Evaluate $\int_a^b f(t)\ dt = F(t)\big|_{t=a}^{t=b} = F(b) - F(a)$.

  c. Compare your results from parts a and b. Since Part II of the FTC holds, the results should be the same. If yours are different, check your calculations.

1. $\int_2^6 5\ dx$

  a. Evaluate the definite integral using a geometric approach.

  b. Evaluate the definite integral using the FTC.

  c. Compare the results.

2. $\int_0^5 (t + 2)\ dt$

  a. Evaluate the definite integral using a geometric approach.

**b.** Evaluate the definite integral using the FTC.

**c.** Compare the results.

**3.** $\int_0^{2\pi} \sin(r)\ dr$

**a.** Evaluate the definite integral using a geometric approach.

**b.** Evaluate the definite integral using the FTC.

**c.** Compare the results.

Based on the results of Task 7-11, Part II of the Fundamental Theorem of Calculus holds for a few examples. Before applying the theorem to more complex functions, consider how Part I can be used to show that the second part always holds.

Suppose $f$ is a continuous function defined on the interval $[a, b]$ and $F$ is any antiderivative of $f$. We want to show that

$$\int_a^b f(t)\ dt = F(b) - F(a).$$

We know from Part I that the accumulation function $A$ generated by $f$ is an antiderivative of $f$. Consequently, since $F$ and $A$ are antiderivatives of the same function, the values of $F$ and $A$ differ by a constant, and there exists a real number of $K$ such that

$$F(x) = A(x) + K \quad \text{for } a \le x \le b.$$

In particular,

$$F(b) = A(b) + K$$

and

$$F(a) = A(a) + K.$$

We will use this relationship to prove Part II, but, first, we need to make two more observations. By definition,

$$A(x) = \int_a^x f(t) \, dt \quad \text{for } a \le x \le b.$$

In particular,

$$A(b) = \int_a^b f(t) \, dt.$$

and

$$A(a) = \int_a^a f(t) \, dt = 0$$

Therefore,

$$F(b) - F(a) = (A(b) + K) - (A(a) + K) \qquad \text{Substitute for } F(b) \text{ and } F(a).$$

$$= \left( \int_a^b f(t) \, dt + K \right) - (0 + K) \qquad \text{Substitute for } A(b) \text{ and } A(a).$$

$$F(b) - F(a) = \int_a^b f(t) \, dt \qquad \text{Simplify.}$$

This is exactly what we wanted to show, so we are done.

As we have mentioned before, Part II of the Fundamental Theorem of Calculus provides you with a straightforward way of calculating the value of a definite integral. For example, in HW5.10, you considered several ways to approximate the value of

$$\int_2^5 (x^2 - 6x + 10) \, dx.$$

Now you can use the theorem to find the exact solution.

$$\int_2^5 (x^2 - 6x + 10) \, dx = (\tfrac{1}{3}x^3 - 3x^2 + 10x)\big|_{x=2}^{x=5}$$

$$= (\tfrac{1}{3}(5^3) - 3(5^2) + 10(5)$$

$$- (\tfrac{1}{3}(2^3) - 3(2^2) + 10(2))$$

$$= \tfrac{50}{3} - \tfrac{32}{3}$$

$$\int_2^5 (x^2 - 6x + 10) \, dx = 6$$

Use this approach to evaluate some definite integrals in Task 7-12.

## Task 7-12: Applying Part II of the FTC

1. Use Part II of the Fundamental Theorem of Calculus to evaluate the following definite integrals.

   **a.** $\int_{1}^{4} 3x^2 \, dx$

   **b.** $\int_{-1}^{2} (5x - x^3) \, dx$

   **c.** $\int_{4}^{16} \left( \frac{1}{\sqrt{t}} - \sqrt{t} \right) dt$

   **d.** $\int_{0}^{1} 6e^x \, dx$

**e.** $\displaystyle\int_0^{2\pi} (\sin(\theta) + \cos(\theta))\, d\theta$

**f.** $\displaystyle\int_1^e \left( t - \frac{1}{t} \right) dt$

**g.** $\displaystyle\int_a^b 4r^3\, dr$

2. For each of the following regions:
   **i.** Sketch the region.
   **ii.** Represent the area of the region by a definite integral.
   **iii.** Find the exact area using Part II of the Fundamental Theorem of Calculus.

   *Note: In HW7.4, part 1, you sketched each of these regions and represented its area by a definite integral.*

**a.** The region bounded by $x = -1$, $x = 1.5$, $y = x^3 + 8$, and the $x$-axis

**b.** The region bounded by $f(t) = \sqrt{t}$, where $1 \le t \le 9$, and the $t$-axis

**c.** The region bounded by the $x$-axis and

$$g(x) = \begin{cases} -x & \text{if } -2 \le x \le -1 \\ -x^2 + 2 & \text{if } -1 < x \le 1 \end{cases}$$

**d.** The region bounded by the graphs of $h(t) = e^t$ and $g(t) = 1$, where $0 \le t \le 1.5$

**e.** The region bounded by $y = x^3$ and $y = 4x$

**f.** The region bounded by $f(x) = \sqrt{x}$, $h(x) = -\sqrt{x}$, and $x = 3$

**g.** The region bounded by $w = x$, $w = x^2$, and $x = 2$

You did it! To recap:

**The Fundamental Theorem of Calculus (FTC).** Let $f$ be a continuous function defined on the interval $[a, b]$.

**Part I:** If $A$ is the accumulation function generated by $f$, then $A$ is an antiderivative of $f$. That is,

$$A'(x) = f(x) \quad \text{for } a \le x \le b.$$

**Part II:** If $F$ is any antiderivative of $f$, then

$$\int_a^b f(x)\ dx = F(x)\Big|_{x=a}^{x=b} = F(b) - F(a).$$

## Unit 7 Homework After Section 3

- Complete the tasks in Section 3. Be prepared to discuss them in class.

- Use the Fundamental Theorem of Calculus to evaluate some definite integrals in HW7.16.

**HW7.16** Evaluate the following definite integrals.

*Note: You may need to do some algebraic manipulations beforehand.*

**1.** $\displaystyle\int_{-2}^{3} (10 + w - 6w^2)\ dw$

**2.** $\displaystyle\int_{-15}^{-5} 10\ dr$

**3.** $\displaystyle\int_{1}^{4} \sqrt{16x}\ dx$

**4.** $\displaystyle\int_{1}^{0} t^2(\sqrt{t} - \sqrt[4]{t})\ dt$

**5.** $\displaystyle\int_{1}^{4} \left(r - \frac{1}{r}\right)^2 dr$

**6.** $\displaystyle\int_{4}^{25} \left(x^{1/2} - \frac{x}{2}\right) dx$

**7.** $\displaystyle\int_{1}^{2} \frac{dx}{x^4}$

**8.** $\displaystyle\int_{-\pi}^{\pi} (2\cos(x) + 6)\ dx$

9. $\int_{-3}^{-1} \frac{2t^5 - t^3}{t^2} \, dt$

11. $\int_{4}^{4} \sqrt[4]{x^4 + 4\sqrt{x} + 4} \, dx$

10. $\int_{2}^{4} \frac{xe^x - 3x^2}{x} \, dx$

12. $\int_{-3}^{1} |w + 1| \, dw$

- Use your CAS to evaluate some definite integrals in HW7.17.

**HW7.17** Put your computer to work.

1. Determine how to evaluate a definite integral using your CAS.

2. Use your CAS to evaluate the definite integrals in Part 1 of Task 7.12 which you calculated using pencil and paper. Your answers should be the same as the ones you derived in Task 7.12 when you did the work by hand. Print a copy of the file containing your work. Label each problem appropriately.

- Explain why the FTC is important in HW7.18.

**HW7.18** Why is the Fundamental Theorem of Calculus fundamental?

The FTC was discovered independently by Sir Isaac Newton (1642–1727) in England and Gottfried Leibniz (1646–1716) in Germany. Because of this discovery, these two mathematicians are said to be the inventors of calculus. Write a one page essay explaining why their discovery was so important.

- Use the Fundamental Theorem of Calculus to solve some problems in HW7.19.

**HW7.19** Solve the following problems using pencil and paper or your CAS.

1. Sketch the region bounded by the given graphs. Find the area of the region.

   a. $y = x^2 + x + 1$, $y = 0$, $1 \le x \le 3$

   b. The $x$-axis and $f(x) = \begin{cases} -x^2 + 8 & \text{if } 0 \le x \le 2 \\ 4 & \text{if } 2 \le x \le 6 \end{cases}$

   c. $y = x^2 + 4$, $y = x$, $x = 0$, $x = 2$

   d. $y = x^2$, $y = 3x$

2. Remember Darcy? In Task 7-1, Darcy hurriedly peddles the bicycle from her meeting to class where her velocity in feet per second is

$$v(t) = 6t - \frac{t^2}{5}.$$

   a. Find the distance Darcy travels in the first 5 seconds—that is, for $0 \le t \le 5$.

   b. Find the distance Darcy travels in 30 seconds.

3. Your new company is doing extremely well, and you estimate that sales will grow continuously at a rate of

$$S'(t) = 3t^2 \text{ thousands of dollars per month.}$$

   **a.** Find your accumulated sales during your first year—that is, for $0 \le t \le 12$.

   **b.** Find your accumulated sales during the second and third months.

   **c.** How many months will it take for your accumulated sales to reach a million dollars?

• Verify some identities and formulas in HW7.20.

**HW7.20**

1. Verify the following identities, by completing the following steps:

   **i.** Evaluate the integral on the left-hand side of the identity using the Fundamental Theorem of Calculus.
   **ii.** Differentiate the result and show that it equals the right-hand side of the identity.

   **a.** $\dfrac{d}{dx}\left( \displaystyle\int_0^x (\sqrt{t} - 2t + 9)\, dt \right) = \sqrt{x} - 2x + 9$ for $x \ge 0$

   **b.** $\dfrac{d}{dx}\left( \displaystyle\int_0^x (t + 4)^2\, dt \right) = (x + 4)^2$

2. Find the following derivatives, where $D_x$ is another notation for the derivative with respect to $x$.

   **a.** $D_x\left( \displaystyle\int_0^x \left( \dfrac{3}{\sqrt{t^2 - 1}} \right)^2 dt \right)$, where $x < -1$ or $x > 1$

   **b.** $D_x\left( \displaystyle\int_0^x t^3 e^t\, dt \right)$

   **c.** $D_x \displaystyle\int_1^x \dfrac{1}{t}\, dt,$ where $x > 0$

3. Verify the following formulas.

   **a.** $\displaystyle\int_a^b \dfrac{(1 - \theta)\cos\theta + \sin\theta}{(1 - \theta)^2}\, d\theta = \dfrac{\sin\theta}{1 - \theta}\Big|_{\theta=a}^{\theta=b}$

   **b.** $16\displaystyle\int_a^b (r^4 - 2r^2)^3(r^3 - r)\, dr = (r^4 - 2r^2)^4\Big|_{r=a}^{r=b}$

   **c.** $\displaystyle\int_a^b \dfrac{-1}{\sqrt{(y^2 + 1)^3}}\, dy = \dfrac{-y}{\sqrt{y^2 + 1}}\Big|_{y=a}^{y=b}$

7

- Prove some properties of Definite Integrals in HW7.21.

**HW7.21** Assume $f$ and $g$ are *integrable* on $[a, b]$—in other words, assume $\int_a^b f(x)\ dx$ and $\int_a^b g(x)\ dx$ exist. For each of the following properties:

   **i.** Use the definition of definite integral to explain why the property holds.

   **ii.** Illustrate the result with a diagram.

**1.** If $f(x) \le M$ for all $x$ in $[a, b]$, then $\int_a^b f(x)\ dx \le M(b - a)$.

**2.** If $m \le f(x)$ for all $x$ in $[a, b]$, then $m(b - a) \le \int_a^b f(x)\ dx$.

**3.** If $f(x) \ge 0$ for all $x$ in $[a, b]$, then $\int_a^b f(x)\ dx \ge 0$.

**4.** If $f(x) \le 0$ for all $x$ in $[a, b]$, then $\int_a^b f(x)\ dx \le 0$.

**5.** If $f(x) \le g(x)$ for all $x$ in $[a, b]$, then $\int_a^b f(x)\ dx \le \int_a^b g(x)\ dx$.

- Examine the Mean Value Theorem for Definite Integrals in HW7.22.

**HW7.22** According to the *Mean Value Theorem for Integrals*:

If $y = f(x)$ is continuous on the closed interval $[a, b]$, then there exists a number $c$ between $a$ and $b$ such that

$$\int_a^b f(x)\ dx = f(c)(b - a).$$

**1.** Examine the geometric interpretation of the Mean Value Theorem for Integrals when $f(x) \ge 0$, for all $x$ in the interval $[a, b]$.

  **a.** Sketch a supportive diagram.

    **(1)** Sketch a pair of axes, and label $a$ and $b$ on the horizontal axis, where $a < b$.

    **(2)** Sketch the graph of a continuous function $f$, where the graph lies above the horizontal axis.

  **b.** Recall that $\int_a^b f(x)\ dx$ equals the area under the graph of $f$ over the interval $[a, b]$. Moreover, if $c$ is a number between $a$ and $b$, then $f(c)(b - a)$ equals the area of the rectangle whose height is $f(c)$ and whose base is the interval $[a, b]$. The Mean Value Theorem for Integrals claims that there exists a value for $c$ so that the areas of these two regions are equal.

    **(1)** On your diagram, label a value for $c$ between $a$ and $b$ such that the area of the region under the graph of $b$ equals the area of the rectangle whose height is $f(c)$.

    **(2)** Label the point $P(c, f(c))$. Shade the associated rectangle.

**2.** Apply the theorem.

    **a.** For each of the following equations:

        **i.** Find a value for $c$ that satisfies the conclusion of the Mean Value Theorem for Integrals.

        **ii.** Sketch the region corresponding to the integral.

        **iii.** Shade the rectangle associated with $c$.

        **(1)** $\displaystyle\int_2^4 (x + 1)\ dx = 8$

        **(2)** $\displaystyle\int_0^3 3x^2\ dx = 27$

    **b.** If $f$ is a constant function, then $c$ can be *any* number between $a$ and $b$. Explain why this is true.

    **c.** Consider the following scenario: Two cars, car 1 and car 2, are traveling on the interstate. One car passes the other, where the velocity of car 1 is 50 mph. During the next 30 seconds, car 1 accelerates to 85 mph, spots a radar detector, and rapidly slows to 50 mph. The driver of car 2 maintains a constant velocity, since she is using her cruise control. The two cars travel the same distance during the 30-second time interval.

        **(1)** Represent the velocity of car 1, over the 30 second interval, by a graph. On the same pair of axes, represent the velocity of car 2. Label the graphs.

        **(2)** Which car is traveling faster when they initially pass?

        **(3)** Label the time(s) when car 1 and car 2 are moving with the same velocity.

        **(4)** The cars pass at $t = 0$. Label the next time that the two cars pass.

- Write your journal entry for this unit. As usual, before you begin to write, review the material in the unit. Think about how it all fits together. Try to identify what, if anything, is still causing you trouble.

**HW7.23** Write your journal entry for Unit 7.

**1.** Reflect on what you have learned in this unit. Describe in your own words the concepts that you studied and what you learned about them. How do they fit together? What concepts were easy? Hard? What were the main (important) ideas? Give some examples of the main ideas.

**2.** Reflect on the learning environment for the course. Describe the aspects of this unit and the learning environment that helped you understand the concepts you studied. What activities did you like? Dislike?

—■—

# Unit 8:

# METHODS OF INTEGRATION

*The definite integral of f represents a busy operation, something undertaken if only in the imagination: <u>down</u> go the partitions on the interval, <u>up</u> go those Mies van der Rohe rectangles, <u>in</u> comes the beetle-browed assessor to compute their areas, the mathematician arriving finally as a great and imperious intellectual Prince sending the sums onward to their appointed limit. It is useful to be reminded that all this busyness results in the end in nothing more than a number, those cloud-capped palaces on various partitions disappearing into thin air when their work is completed.*

David Berlinski, *A Tour of the Calculus*, p. 283, Pantheon Books, NY, 1995.

## OBJECTIVES

1. Integrate functions using:

   • The method of substitution
   • Integration by parts
   • The table lookup approach

2. Use numerical integration—in particular, the Trapezoidal Rule—to approximate:

   • The definite integral of a function whose antiderivative does not have an elementary form
   • The definite integral associated with a discrete function

3. Approximate the definite integral associated with a discrete function by fitting a curve to the data and integrating the fit function.

In this unit, you will learn new strategies for evaluating integrals and deepen your understanding of what an integral represents mathematically.

In order to be able to use the Fundamental Theorem of Calculus, you need to know how to find an antiderivative of the underlying function. At this point, you know how to find antiderivatives of functions that are sums of scalar multiples of the basic functions, $x^n$, $1/x$, $e^x$, $\sin(x)$ and $\cos(x)$. For instance, you know how to find

$$\int\left(\frac{x^3}{3} - \frac{3}{x^3} + \sqrt[3]{x}\right)dx, \quad \int\left(0.75e^r + \frac{2}{r}\right)dr,$$

and

$$\int\left(\pi\cos(t) + \tfrac{3}{4}\sin(t)\right)dt.$$

But what about finding an antiderivative of a function that cannot be expressed as a linear combination of the basic functions, such as

$$\int(2x - 3)\sqrt{x^2 - 3x + 6}\,dx \quad \text{or} \quad \int x\cos(x)\,dx?$$

To solve these types of problems, you need to develop more sophisticated integration techniques. There are many different techniques, but in this unit, you will learn two: integration by substitution and integration by parts. You will also learn how to use integration tables to look up integrals.

There are situations, however, where none of the techniques applies. In particular:

- The integration techniques do not apply if the function is represented by an expression, but there does not exist an *elementary form* for its antiderivative. For example, the antiderivative

$$f(x) = e^{-x^2/2},$$

   which occurs in many probability applications, does not have an elementary form.
- The integration techniques do not apply if a function is represented by a data set—or a collection of ordered pairs—but not by an expression. For example, the region bounded by the Susquehanna River, which you examined in Unit 5, is represented by a collection of ordered pairs, but not by an expression.

Both of these types of situations can be handled by approximating the value of the definite integral using *numerical integration*. This involves covering the region defined by the function with shapes with which you are familiar (rectangles, trapezoids, and so on), and then using them to approximate the value of the desired definite integral. This, of course, is the approach you used in Unit 5 to develop the definition of definite integral. In this unit, you will look more carefully at the trapezoidal approach, because it usually gives a more precise approximation than the rectangular approach, with the same number of subintervals.

In addition to numerical integration, another way to approximate the definite integral associated with a set of data points is to fit a curve to the data, or maybe several curves to different pieces of the data, and then use the fit function to approximate the definite integral. This approach is feasible if the graph of the data is fairly regular.

Integration is useful for finding the areas of odd-shaped regions or figuring out how quantities accumulate. Moreover, solutions to integrals are extraordinarily powerful in helping social and natural scientists find functional relationships that can be used to describe the behavior of systems mathematically. Techniques for evaluating definite integrals enable those in the applied sciences, such as engineers, meteorologists, and economists, to use theoretical understanding of systems to perform the calculations needed to design bridges, forecast the weather, or make projections of the national debt. In this unit, in addition to learning new strategies, you will use them to investigate some real-life situations.

## SECTION 1

### Integrating by Substitution

One way to evaluate a complicated integral—which you do not know how to solve—is to transform the given integral into an equivalent integral—which you do know how to solve. You used this approach in the last unit when you used algebraic manipulation to simplify a given expression before finding its antiderivative. For example,

$$\int \frac{2x^{2/3} - 6x^{5/3}}{x} = \int (2x^{-1/3} - 6x^{2/3}) \ dx = 3x^{2/3} - \frac{18}{5}x^{5/3} + C.$$

The technique of *integration by substitution* also uses this approach, but in this case, instead of using algebraic manipulation to transform the given integral into an equivalent one, you use substitution or a *change of variables.*

Before looking at how substitution works in general, consider a particular situation. As you know, regions with different shapes can have the same area. For example, a square with 4-foot sides has the same area as a $2 \times 8$-foot rectangle or a parallelogram which is 1 foot high and 16 feet long. You can show that these regions have the same area by using the area formulas for regular shapes. But how might you show that two regions, which are bounded by two different continuous functions, have the same area? The goal of the next task is for you to answer this question, first by convincing yourself that two given regions have the same area using paper and scissors, and then by proving that the two regions have the same area by using the new technique—integration by substitution.

## Task 8-1: Equating Areas

Consider the functions

$$h(x) = x\sqrt{x^2 + 16}, \quad \text{where } 0 \le x \le 3,$$

$$g(u) = \frac{\sqrt{u}}{2}, \quad \text{where } 16 \le u \le 25.$$

Obviously, $h$ and $g$ are different functions, and when you graph $h$ and $g$, you will see that the shapes of their graphs are quite different. However, it turns out that the regions bounded by the two graphs and the horizontal axis have the same area. The goal of this task is to use a graphical approach to convince yourself that the areas are the same, in which case the definite integrals representing the areas will have the same value.

1. First try a graphical approach. If two regions have the same area, then one region can be cut into pieces which exactly cover the other region. Try this approach for the regions bounded by the graphs of $h$ and $g$.

   a. Carefully graph $h$ and $g$ on the graph paper on the next two pages.

   b. Outline the regions bounded by the two graphs and the horizontal axis.

   c. Shade the region bounded by the graph of $g$ and cut out the region.

   d. Cover the region bounded by $h$ with the region bounded by $g$ by cutting the region bounded by $g$ into pieces and taping the pieces of $g$ onto the region determined by $h$. Since the two regions have the same area, the pieces from the region determined by $g$ should more or less cover the region determined by $h$. Do they?

2. Next try a calculus approach. If two regions have the same area, then the definite integrals representing the regions have the same value. Consider this approach for the regions bounded by the graphs of $h$ and $g$.

   a. Express the area of each region in terms of a definite integral.

   b. You can solve one of the definite integrals using the integration methods you have learned to date, but not the other. Calculate the area of the region by evaluating the definite integral that you know how to solve.

   c. Explain why the other integral is difficult to solve.

Graph of $h(x) = x\sqrt{x^2 + 16}$, where $0 \le x \le 3$

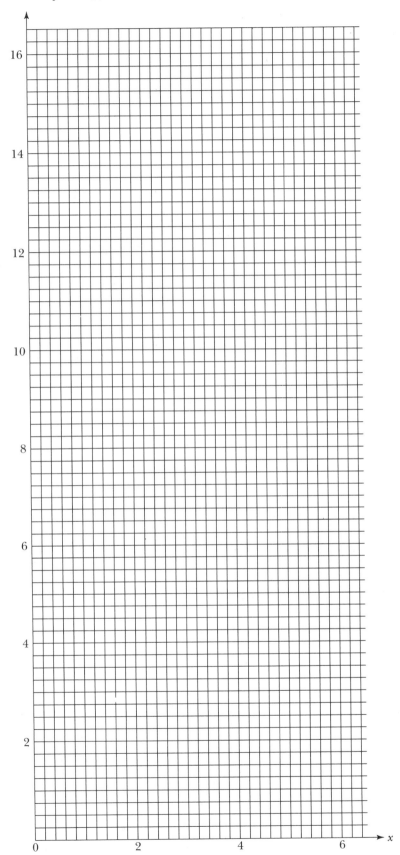

Graph of $g(u) = \dfrac{\sqrt{u}}{2}$, where $16 \le u \le 25$

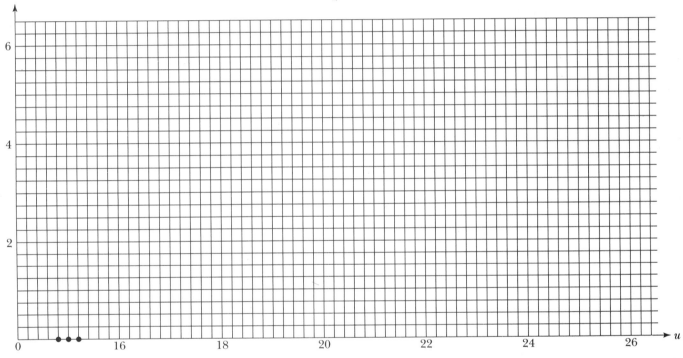

Based on the cutting and pasting activities you did in the last task, it appears that the area of the region bounded the graph of $h$ equals the area bounded by $g$. If this is the case, then

$$\int_0^3 h(x)\ dx = \int_{16}^{25} g(u)\ du$$

or

$$\int_0^3 x\sqrt{x^2 + 16}\ dx = \int_{16}^{25} \frac{\sqrt{u}}{2}\ du$$

since these integrals represent the areas of the two equivalent regions.

What makes this equality so important is that the left-hand side is a complicated integral which you *do not* know how to evaluate, whereas the right-hand side is a basic integral which you *do* know how to evaluate. Therefore, if you knew how to transform the complicated one (involving the $x$'s) into the basic one (involving the $u$'s), you could find the value of the complicated one by evaluating the basic. One way to do this is to use the change-of-variables technique or integration by substitution. Let us examine how this technique applies to this example.

Because both integrals involve taking a square root, namely $\sqrt{x^2 + 16}$ and $\sqrt{u}$, it seems reasonable to let

$$u = x^2 + 16.$$

Let us try to use this substitution to transform the integral on the left—which is in terms of $x$—into the one on the right—which is in terms of $u$.

First, how does this substitution affect the limits of integration? Since $x$ varies from 0 to 3, $u$ varies from 16 to 25—that is, when $x = 0$,

$$u = 0^2 + 16 = 16,$$

and when $x = 3$,

$$u = 3^2 + 16 = 25$$

or

| x | u |
|---|---|
| 0 | 16 |
| 3 | 25 |

Substituting $u$ for $x^2 + 16$, 16 for 0, and 25 for 3 in the original integral on the left gives

$$\int_0^3 x\sqrt{x^2 + 16}\ dx = \int_{16}^{25} x\sqrt{u}\ dx.$$

You want the right side to be strictly in terms of $u$—with no $x$'s—but you still have some $x$'s on the right side, namely $x$ and $dx$. How does the change of variables affect $dx$? By assumption,

$$u = x^2 + 16.$$

If you take the derivative of $u$ with respect to $x$, then

$$\frac{du}{dx} = 2x.$$

Therefore,

$$du = 2x\,dx \quad \text{or} \quad \frac{1}{2}du = x\,dx.$$

This result enables you to transform the complicated integral with the $x$'s into a basic one in terms of $u$:

$$\int_0^3 x\sqrt{x^2 + 16}\,dx = \int_0^3 \sqrt{x^2 + 16}\,x\,dx \qquad \text{Rearrange terms.}$$

$$= \int_{16}^{25} \sqrt{u}\,\frac{1}{2}\,du \qquad \text{Substitute } u \text{ for } x^2 + 16$$
$$\text{and } \frac{1}{2}du \text{ for } x\,dx. \text{ Change}$$
$$\text{the limits of integration.}$$

$$\int_0^3 x\sqrt{x^2 + 16}\,dx = \int_{16}^{25} \frac{\sqrt{u}}{2}\,du \qquad \text{Rearrange terms.}$$

Because the two integrals are equivalent, you can find the value of $\int_0^3 x\sqrt{x^2 + 16}\,dx$ by evaluating $\int_{16}^{25} \frac{\sqrt{u}}{2}\,du$ instead. Consequently, based on the work you did in the last task,

$$\int_0^3 x\sqrt{x^2 + 16}\,dx = \int_{16}^{25} \frac{\sqrt{u}}{2}\,du = \frac{61}{3}.$$

Let us summarize what just happened. First, we defined $u = f(x)$—in this case, $u = x^2 + 16$. We used this expression to find the new limits of integration—in this case,

| x | u |
|---|---|
| 0 | 16 |
| 3 | 25 |

and we used the expression to find $du$ in terms of $x$ and $dx$—in this case, $\frac{du}{dx} = 2x$, so $\frac{1}{2}\,du = x\,dx$. Next, we used substitution to replace all the $x$'s with $u$'s in the given integral, which led to

$$\int_0^3 x\sqrt{x^2 + 16}\,dx = \int_{16}^{25} \frac{\sqrt{u}}{2}\,du.$$

Finally, we evaluated the new definite integral (which is equivalent to the given one and is one that you know how to solve).

We did pull one bit of hocus-pocus as we went through all this, when we claimed that $du = 2x\,dx$ since $\frac{du}{dx} = 2x$. Let us think about this in general and see why this is an acceptable thing to do. Consider the diagram:

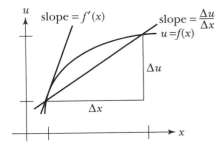

Observe that

$$\text{the slope of the secant line} = \frac{\Delta u}{\Delta x}$$

and

$$\text{the slope of the tangent line} = f'(x).$$

Now, the slope of the secant line approximates the slope of the tangent line when $\Delta x$ is small—that is,

$$\frac{\Delta u}{\Delta x} \approx f'(x).$$

So

$$\Delta u \approx f'(x)\Delta x.$$

As $\Delta x \to 0$, the approximation gets better and better, since the slope of the secant line gets closer and closer to the slope of the tangent line. Therefore, since $dx$ represents an infinitesimally small $\Delta x$ and $du$ represents an infinitesimally small $\Delta u$, it makes sense to say that

$$du = f'(x)\ dx,$$

where $du$ and $dx$ are called *differentials*.

---

## Task 8-2: Examining the Strategy Underlying Substitution

**1.** Before evaluating some actual integrals, find some differentials, noting that if

$$u = f(x),$$

then

$$\frac{du}{dx} = f'(x)$$

and

$$du = f'(x)\ dx.$$

For example, if

$$u = x^2 + 16,$$

then

$$\frac{du}{dx} = 2x$$

and

$$du = 2x \, dx \quad \text{or} \quad x \, dx = \frac{1}{2} \, du = \frac{du}{2}.$$

**a.** Consider $u = 7x - 6$. Find the following:

(1) $\dfrac{du}{dx}$

(2) $du$

(3) $dx$

**b.** Consider $u = -\dfrac{x^3}{4}$. Find the following:

(1) $\dfrac{du}{dx}$

(2) $du$

(3) $x^2 \, dx$

**c.** Consider $w = x^2 + 2x - 9$. Find the following:

(1) $\dfrac{dw}{dx}$

(2) $dw$

(3) $(x + 1) \, dx$

**d.** Consider $u = \ln(8t)$. Find the following:

(1) $\dfrac{du}{dt}$

(2) $du$

**e.** Consider $w = kx$, where $k$ is a nonzero constant. Find the following:

(1) $\dfrac{dw}{dx}$

(2) $dw$

(3) $dx$

2. Use substitution to show that the definite integrals given below in parts a to c have the indicated values. Show all your work.

As you complete the exercises, keep in mind that the goal is to transform the given integral, which you do not know how to evaluate, into an integral which you can evaluate. In general, this involves the following:

   **i.** Define $u = f(x)$. (Note that in the following problems, we have provided you with an appropriate choice for $u$. In future exercises, you will have to choose $u$. As you evaluate each of these integrals, try to determine why we have selected the given $u$.)

   **ii.** Find the new limits of integration.

   **iii.** Express $du$ in terms of $x$ and $dx$, noting that $du = f'(x)\, dx$.

   **iv.** Use the information from steps i through iii to express the given integral in terms of $u$; that is, use substitution to replace all the $x$'s ($t$'s, $\theta$'s, etc.) with $u$'s.

   **v.** Evaluate the new definite integral (which is equivalent to the given one and should be one you can do).

Before beginning these exercises, it may be helpful to review the example which you considered before this task.

**a.** Use the method of substitution to show that

$$\int_0^1 \frac{2x}{(x^2 + 1)^5}\, dx = \frac{15}{64}.$$

   **i.** Let $u = x^2 + 1$.

   **ii.**

**b.** Use the method of substitution to show that

$$\int_2^5 \frac{2}{2t - 3} \, dt = \ln(7).$$

**i.** Let $u = 2t - 3$.

**ii.**

**c.** Use the method of substitution to show that

$$-\int_{\pi/8}^{\pi/2} \sin(4\theta) \, d\theta = \frac{1}{4}.$$

**i.** Let $u = 4\theta$.

**ii.**

**3.** In the last part, you used substitution to evaluate some definite integrals. You can also use substitution to find antiderivatives—for instance, to find

$$\int \frac{2x}{(x^2 + 1)^5} \, dx$$

you can let $u = x^2 + 1$, just as you did in part 2a, then $du = 2x \, dx$ and

$$\int \frac{2x}{(x^2 + 1)^5} \, dx = \int \frac{1}{u^5} \, du \qquad \text{Substitute } u = x^2 + 1 \\ \text{and } du = 2x \, dx.$$

$$= \int u^{-5} \, du \qquad \text{Simplify.}$$

$$= \frac{u^{-4}}{-4} + C \qquad \text{Integrate.}$$

$$\int \frac{2x}{(x^2 + 1)^5} \, dx = \frac{(x^2 + 1)^{-4}}{-4} + C \qquad \text{Substitute } x^2 + 1 \text{ back} \\ \text{in for } u.$$

Note that the difference between using substitution to evaluate a definite integral and using substitution to find a general antiderivative is that in the case of a definite integral you find new limits of integration and calculate a specific value using these new limits. In the case of finding an antiderivative, there are no limits of integration, so it makes no sense to do this. You do, however, want your answer to be in terms of the *same variable* as the original problem, which means that after evaluating the equivalent integral (in terms of $u$), you need to substitute back in, which in this case meant replacing the $u$ with $x^2 + 1$. Then, as you have done in the past, you can check that the result is correct by findings its derivative—that is, you can show that

$$\frac{d}{dx} \left( \frac{(x^2 + 1)^{-4}}{-4} + C \right) = \frac{2x}{(x^2 + 1)^5}.$$

Use substitution to find the antiderivatives in parts a to c below. Show all your work and use differentiation to check your answer. In general, this involves the following:

   **i.** Define $u = f(x)$. Again, we have provided you with an appropriate choice for $u$. Try to determine why we have selected the given $u$.

   **ii.** Express $du$ in terms of $x$ and $dx$, noting that $du = f'(x) \, dx$.

   **iii.** Use the information from steps i and ii to express the given integral in terms of $u$.

   **iv.** Find the antiderivative (which is equivalent to the given one and should be one you can do).

   **v.** Express the result in terms of the same variable as the original antiderivative.

   **vi.** Use differentiation to check that your answer is correct.

 **a.** Use the method of substitution to show that

$$\int 2(2x - 4)^4 \, dx = \frac{1}{5}(2x - 4)^5 + C.$$

   **i.** Let $u = 2x - 4$.

**ii.**

**b.** Use the method of substitution to show that

$$\int xe^{x^2}\,dx = \frac{1}{2}e^{x^2} + C.$$

**i.** Let $u = x^2$.

**ii.**

**c.** Use the method of substitution to show that

$$2\int t\cos(t^2 + \frac{\pi}{2})\,dt = \sin(t^2 + \frac{\pi}{2}) + C.$$

**i.** Let $u = t^2 + \frac{\pi}{2}$.

**ii.**

At this point, you can use substitution when we provide you with an appropriate choice for $u$. But, why did we pick the given $u$? The goal of the next task is to develop a general approach for finding antiderivatives using substitution and to help you begin to recognize appropriate choices for $u$.

## Task 8-3: Inspecting Situations Where Substitution Applies

**1.** The main idea underlying substitution is to transform a given integral into a familiar form. First recall five rules for finding antiderivatives of basic functions, as you by fill in the table given below.

| Antiderivative $\int g(u)\, du$ | | | | | |
|---|---|---|---|---|---|
| Basic Function $g(u)$ | $u^n,\ n \neq -1$ | $\dfrac{1}{u},\ u \neq 0$ | $e^u$ | $\sin(u)$ | $\cos(u)$ |

**2.** Consider some situations where substitution applies. For each of the five general cases given below in parts a through e:

  **i.** Use substitution to find the antiderivative in the general case.

  - Express the given integral in terms of $u$ by substituting $u = f(x)$ and $du = f'(x)\ dx$.
  - Evaluate the "new" integral by using rules for integrating basic functions (see part 1 above).
  - Express the solution in terms of $x$ by substituting $f(x)$ back in for $u$.

  **ii.** Make up an example that has the same form.

  - Give an example of an antiderivative that has the same form as the general case.
  - Evaluate your antiderivative by using substitution.
  - Check that your answer is correct by differentiating the result.

*Note: To help you get started, we have completed the first problem. Read through it carefully before continuing.*

**a.** Consider $\int (f(x))^n f'(x) \, dx$, where $n \neq -1$.

   **i.** Use substitution to find the antiderivative of this general case.

      Let $u = f(x)$. Then $du = f'(x) \, dx$ and

$$\int (f(x))^n f'(x) \, dx = \int u^n \, du \qquad \text{Substitute } u = f(x) \\ \text{and } du = f'(x) \, dx.$$

$$= \frac{u^{n+1}}{n+1} + C \qquad \text{Integrate the basic} \\ \text{function } u^n,\ n \neq -1.$$

$$\int (f(x))^n f'(x) \, dx = \frac{(f(x))^{n+1}}{n+1} + C \qquad \text{Substitute } f(x) \text{ for } u.$$

  **ii.** Give an example that has the same form as the general case. Find its antiderivative and check your solution.

     $\int (x^2 + 6)^3 2x \, dx$ has the form $\int (f(x))^n f'(x) \, dx$, where $f(x) = x^2 + 6$.

     To find the antiderivative, let $u = x^2 + 6$. Then $du = 2x \, dx$ and

$$\int (x^2 + 6)^3 2x \, dx = \int u^3 \, du \qquad \text{Substitute } u = x^2 + 6 \\ \text{and } du = 2x \, dx.$$

$$= \frac{u^4}{4} + C \qquad \text{Integrate the basic} \\ \text{function } u^3.$$

$$\int (x^2 + 6)^3 2x \, dx = \frac{1}{4}(x^2 + 6)^4 + C \qquad \text{Substitute } x^2 + 6 \text{ for } u.$$

     Check the solution using differentiation:

$$\frac{d}{dx}\left(\frac{1}{4}(x^2 + 6)^4 + C\right) = (x^2 + 6)^3 \frac{d}{dx}(x^2 + 6) = (x^2 + 6)^3 2x.$$

**b.** Consider $\int \frac{1}{f(x)} f'(x) \, dx$, where $f(x) \neq 0$.

   **i.** Use substitution to find the antiderivative of this general case.

  **ii.** Give an example that has the same form as the general case. Find its antiderivative and check your solution.

**c.** Consider $\int e^{f(x)} f'(x)\,dx$.

   **i.** Use substitution to find the antiderivative of this general case.

   **ii.** Give an example that has the same form as the general case. Find its antiderivative and check your solution.

**d.** Consider $\int \sin(f(x)) f'(x)\,dx$.

   **i.** Use substitution to find the antiderivative of this general case.

   **ii.** Give an example that has the same form as the general case. Find its antiderivative and check your solution.

e. Consider $\int \cos(f(x))f'(x)\,dx$.

i. Use substitution to find the antiderivative of this general case.

ii. Give an example that has the same form as the general case. Find its antiderivative and check your solution.

3. Show that the following integrals are equivalent. Identify the general form and then:

i. Define $u$.

ii. Find $du$.

iii. Show that the integral on the left can be transformed into the integral on the right by substituting $u$'s for all the $x$'s ($t$'s, $\theta$'s, etc.). In the case of a definite integral, show that limits of integration with respect to $u$ are correct.

a. $\int \left(\dfrac{1}{x^2} - 1\right)^{100} x^{-3}\,dx = -\dfrac{1}{2}\int u^{100}\,du$

**b.** $\displaystyle\int_0^{\sqrt{\pi}} t\sin(t^2)\ dt = \frac{1}{2}\int_0^{\pi} \sin(u)\ du$

**c.** $\displaystyle\int_{-2}^{1} x^3 e^{x^4-1}\ dx = \frac{1}{4}\int_{15}^{0} e^u\ du$

**d.** $\displaystyle\int \frac{t^2}{t^3-4}\ dt = \frac{1}{3}\int \frac{1}{u}\ du$

**e.** $-\int \dfrac{\cos(1/t)}{t^2}\, dt = \int \cos(u)\, du$

Since differentiation and integration are inverse processes, every rule for differentiation can be transformed into a corresponding rule for integration. For instance, in the last task, you used the method of substitution to transform the Extended Power Rule

$$\frac{d}{dx}\left(\frac{1}{n+1}\,(f(x))^{n+1}\right) = (f(x))^n f'(x), \quad \text{provided } n \neq -1,$$

into the rule for integrating functions raised to a power

$$\int (f(x))^n f'(x)\, dx = \frac{1}{n+1}\,(f(x))^{n+1} + C$$

by reversing the process of differentiation. Similarly, you used the method of substitution to transform the rule for differentiating logarithmic functions

$$\frac{d}{dx}\left(\ln|f(x)|\right) = \frac{f'(x)}{f(x)}, \quad \text{provided } f(x) \neq 0,$$

into the integration rule

$$\int \frac{f'(x)}{f(x)} \, dx = \ln|f(x)| + C;$$

you transformed the rule for differentiating exponential functions

$$\frac{d}{dx}(e^{f(x)}) = e^{f(x)}f'(x)$$

into the integration rule

$$\int e^{f(x)}f'(x) \, dx = e^{f(x)} + C;$$

and you transformed the rules for differentiating trigonometric functions

$$\frac{d}{dx}(\sin(f(x))) = \cos(f(x))f'(x)$$

and

$$\frac{d}{dx}(-\cos(f(x))) = \sin(f(x))f'(x)$$

into rules for integrating trigonometric functions

$$\int \cos(f(x))f'(x) \, dx = \sin(f(x)) + C$$

and

$$\int \sin(f(x))f'(x) \, dx = -\cos(f(x)) + C.$$

Each of the differentiation rules given above was developed by applying the Chain Rule. So, it is probably not surprising that, in general, the Chain Rule can be transformed into integration by substitution. In particular, if $G$ is any antiderivative of $g$, then $G' = g$ and by the Chain Rule

$$\frac{d}{dx}(G(f(x))) = G'(f(x))f'(x) = g(f(x))f'(x).$$

Reversing the process of differentiation gives you

$$\int g(f(x))f'(x) \, dx = G(f(x)) + C, \quad \text{where } G' = g,$$

or if you let $u = f(x)$, then $du = f'(x) \, dx$ and use integration by substitution

$$\int g(f(x))f'(x) \, dx = \int g(u) \, du = G(u) + C = G(f(x)) + C.$$

The trick to using integration by substitution is to be able to recognize when the function that you are integrating—that is, when the *integrand*— has the form $g(f(x))f'(x)$. In this case, if you substitute $u = f(x)$ and $du = f'(x) \, dx$, then the integral reduces to the equivalent integral $\int g(u) \, du$.

Recognizing when to use integration by substitution takes lots of practice. The next task will help you become more proficient.

## Task 8-4: Using Substitution

1. Use substitution to evaluate the antiderivatives given below in parts a to d.

   - Show all your work, clearly stating your choice for $u$. (See list of steps in Task 8-2, part 3.)
   - Remember to express your final answer in terms of the same variable as the original problem.
   - Check your answer by differentiating.

   a. $\int \dfrac{x^2 + 1}{(x^3 + 3x - 1)^6} \, dx$

   Check:

   b. $\int \dfrac{\cos(\sqrt{r})}{\sqrt{r}} \, dr$

   Check:

   c. $\int e^{\sin(x)} \cos(x) \, dx$

   Check:

**d.** $\int \dfrac{p}{p^2 - 1}\, dp$

Check:

2. Use substitution to evaluate the definite integrals in part a to c.

   - Show all your work, clearly stating your choice for $u$. (See list of steps in Task 8-2, part 2.)
   - When you use substitution to evaluate a definite integral, remember to change the limits of integration by finding the values for $u$ that correspond to the given values for $x$, and then evaluate the antiderivative of $g(u)$ at the new limits of integration. You do not have to express the antiderivative in terms of the original variable. However, if you do, then you must use the original limits of integration to calculate the value of the definite integral. This is usually much more work.

   **a.** $\displaystyle\int_{-1}^{2} (x^2 - 4)^3 x\, dx$

**b.** $\displaystyle\int_{1}^{4} \frac{2n+1}{n^2+n-1}\, dn$

**c.** $\displaystyle\int_{\pi}^{3\pi/2} \sin(t-\pi)\, dt$

In the last task in this section, you will tackle a project where you need to use integration by substitution. The question you wish to answer is: Approximately how many humans have lived since 1650?

Conventional wisdom says that a healthy population will tend to grow at a rate which is proportional to its size. Between A.D. 1650 and 1987, the human population grew at a rate which is faster than this. In fact, historical data between these dates can be described fairly well by a function of the form

$$P(t) = \frac{1}{k(T-t)},$$

where $P$ is the population in year $t$. The growth described by this equation is faster than exponential growth. This is a rather amazing equation because most population biologists predict that populations increase exponentially until they encounter shortages of necessary resources such as food or water. Once shortages are encountered, then biologists believe that the growth rate of the population will slow and eventually level off.

If the function claimed to model the data seems reasonable and you assume it can be extrapolated to describe population levels before A.D. 1650 and after A.D. 1987, then you can make some interesting speculations. For example, you can approximate the total number of people who have lived between the year 1650 and now. To do this, you will need to estimate how long an average person lives and integrate a definite integral using the method of substitution.

## Task 8-5: (Project) Tracking the Human Race

The number of humans estimated to be on the Earth at various times between 1650 and 1987 are summarized in Table 1.

By using mathematical modeling, mathematicians have made the claim that these data can be described by the equation

$$P(t) = \frac{1}{k(T - t)},$$

where $P(t)$ represents the estimated population in billions of people as a function of the year A.D., $k = 5.06 \times 10^{-3}$ and $T = 2026.5$ years A.D.*

In the following parts, show that $P$ is a reasonable model; then use the function to estimate how many people have lived since 1650, and use it to estimate how long it will take before there are an infinite number of people on the Earth.

Table 1. World Population Data

| $t$ (year A.D.) | $P$ (billions) |
|---|---|
| 1650 | 0.55 |
| 1750 | 0.73 |
| 1800 | 0.91 |
| 1850 | 1.17 |
| 1900 | 1.61 |
| 1920 | 1.83 |
| 1930 | 2.07 |
| 1940 | 2.30 |
| 1950 | 2.52 |
| 1960 | 3.01 |
| 1976 | 4.00 |
| 1987 | 5.00 |

1. Plot the population data given in Table 1 and the function $P$ from 1650 to 2000, on the same pair of axes, using your CAS. Print a copy of your graph and place it below.

2. How well does the function $P$ fit the population data?

---

*This equation has been adapted from one which is just slightly more complicated. D. A. Smith, "Human Population Growth: Stability or Explosion?" *Mathematics Magazine*, 50 (1977), 186–197, and D.H. Hallet and D.A. Smith, notes for "Workshop 12: AIDS, Airline Collisions, and Population Trends" for the *National Science Foundation Conference on Excellence in Mathematics and Science Education K–16*. Feb. 9–11, 1992.

3. Recent data estimate that the world population was 5.32 billion people in 1990.*

   a. Use the function $P$ to predict the world population for 1987 and 1990 and enter your predictions in the table below.

   | Year A.D. $t$ | Billions pop. data | Billions $P$ predicted |
   |---|---|---|
   | 1987 | 5.00 | |
   | 1990 | 5.32 | |

   b. Compare the population data with the values predicted by $P$ for 1990. Is the function $P$ a good predictor for the population?

4. How far in the future might the function $P$ work? When the denominator of the population function $P$ becomes zero, the population goes to infinity. The year that this happens is called the Doomsday Year. What year is this? (Let us hope this function only works between 1650 and 1987!)

5. Let $L$ be the average life span. Assume that the value of $L$ has not changed a lot since 1650. What do *you* estimate the human life span to be?

$$L = \underline{\hspace{1.5cm}} \text{ years}$$

6. Explain why you can use integration to estimate the number of people who have lived.

7. You can use integration to estimate the number of people $N$ who have lived in the world since 1650, but you need to divide your result by the average life span. Why?

---

*World Population, July 1990: 5,316,644,000, estimated growth rate 1.7%. (Source CIA World Factbook 1991.)

8. The estimated number of people who have lived from 1650 until now is given by the integral

$$N = \frac{1}{L} \int_{t_i}^{t_f} P(t) \, dt$$

where $L$ is the average human life span (which you estimated in part 5).

**a.** Find the limits of integration.

**b.** Use substitution to calculate the value of $N$.

## Unit 8 Homework After Section 1

- Complete the tasks in Section 1. Be prepared to discuss them in class.

- Use substitution to find some antiderivatives and solve some definite integrals in HW8.1.

**HW8.1** Evaluate the following integrals. Show all your work. Check the antiderivatives by differentiating.

1. $\int (3x - 1)^9 \, dx$

2. $\int_0^{\pi/4} \cos(2t + \pi) \, dt$

3. $\int \dfrac{x + 1}{x^2 + 2x - 9} \, dx$

4. $\int \dfrac{(\ln(7t))^{10}}{t \, dt}$

5. $\int \dfrac{1}{x^2} \sqrt[5]{\dfrac{1}{x} + 5} \, dx$

6. $\int_0^1 e^x \cos(\pi e^x) \, dx$

7. $\int_{-1}^2 \dfrac{r^2 - 1}{(r^3 - 3r + 3)^3} \, dr$

8. $\int \sin^7(\theta) \cos(\theta) \, d\theta$

9. $\int \dfrac{1}{\sqrt{x}} \cos(\sqrt{x}) \, dx$

10. $\int_0^{\pi/4} (\cos(2r) + \sin(2r)) \, dr$

11. $\int \dfrac{6}{\sqrt{4 - 3q}} \, dq$

12. $\int x e^{x^2 + 3} \, dx$

13. $\int_0^1 (p^4 - 2p^2 + 3)^{10} (p^3 - p) \, dp$

14. $\int \dfrac{(\sqrt{w} + 6)^5}{\sqrt{w}} \, dw$

15. $\int_1^{14} \sqrt[3]{2z - 1} \, dz$

16. $\int \left(1 + \dfrac{1}{m}\right)^{-3} \left(\dfrac{1}{m^2}\right) \, dm$

17. $\int_{3\pi/2}^{2\pi} \dfrac{\sin(x)}{\sqrt{1 + \cos(x)}} \, dx$

18. $\int_1^4 \dfrac{1}{(\sqrt{k} + 1)^3 \sqrt{k}} \, dk$

**19.** $\int_1^2 \frac{e^{2/t}}{t^2}\, dt$ **20.** $\int \frac{v^2 - 1}{v^3 - 3v + 3}\, dv$

- Use substitution to find the area of a region in HW8.2.

**HW8.2** Consider the region under the graph of $f(x) = 3\sin(x/2)$ over the closed interval $[0, 2\pi]$.

1. Sketch the region.

2. Use substitution to find the area of the region.

3. When you use substitution, the integral representing the area of the given region reduces to an integral that represents the area of a new region. Sketch the new region.

4. What can you say about the two regions you sketched in parts 1 and 3?

- Use substitution to analyze some "real" situations in HW8.3.

**HW8.3** Analyze the following situations.*

1. The U.S. divorce rate is approximated by

$$D(t) = 100{,}000e^{0.025t},$$

   where $D(t)$ is the number of divorces occurring at time $t$, where $t$ is the number of years since 1900—that is, $t = 0$ corresponds to the beginning of 1900.

   **a.** Find the total number of divorces from the beginning of 1900 to the beginning of 1993.

   **b.** Find the total number of divorces from the beginning of 1990 through the end of 1992.

2. A roller coaster is made in such a way that it is $y$ meters above the ground and $x$ meters from the starting point, where

$$y = 15 + 15\sin\left(\tfrac{\pi}{50}x\right).$$

   Find the area under the roller coaster from the starting point to the point 100 meters away.

- Use substitution to find the rule for integrating the general exponential function in HW8.4. (Note: See relevant material in Task 6-16 and HW7.13.)

**HW8.4** If $a$ is a positive number, you know that

$$\frac{d}{dx}\left(a^{f(x)}\right) = (\ln a)\, a^{f(x)} f'(x)$$

(see Task 6-16). Moreover, since $a$ is fixed, $\ln a$ is constant and

$$\frac{d}{dx}\left(\frac{1}{\ln a}\, a^{f(x)}\right) = a^{f(x)} f'(x).$$

---

*See *Applied Calculus* by Bittenger and Morrell, Addison-Wesley, New York, 1993.

This differentiation rule transforms into the integration rule

$$\int a^{f(x)} f'(x)\ dx = \frac{1}{\ln a} a^{f(x)} + C;$$

or, if $u = f(x)$, then $du = f'(x)\ (dx)$ and

$$\int a^u\ du = \frac{1}{\ln a} a^u + C.$$

Evaluate the following integrals, using the method of substitution in conjunction with the new rule:

**1.** $\int 10^{5x}\ dx$

**5.** $\int_{-4}^{-2} \frac{1}{2^t}\ dt$

**2.** $\int 5^{-10t}\ dt$

**6.** $\int_1^2 \frac{4^{\sqrt{x}}}{\sqrt{x}}\ dx$

**3.** $\int r^2 3^{r^3}\ dr$

**7.** $\int \frac{10^{\ln(x)}}{x}\ dx$

**4.** $\int 2^{(x+1)^2}(x + 1)\ dx$

**8.** $\int 4^{4x-1}\ dx$

- Use substitution to find antiderivatives of other trigonometric functions in HW8.5. (See related homework exercises HW6.16 and HW7.14.)

**HW8.5**

**1.** You know that if $f$ is differentiable, then the general rules for differentiating sine and cosine

$$\frac{d}{dx}(\sin(f(x))) = \cos(f(x))f'(x), \qquad \frac{d}{dx}(-\cos(f(x))) = \sin(f(x))f'(x)$$

can be transformed into rules for integrating sine and cosine

$$\int \cos(f(x))f'(x)\ dx = \sin(f(x)) + C,$$

$$\int \sin(f(x))f'(x)\ dx = -\cos(f(x)) + C;$$

or, if $u = f(x)$, then $du = f'(x)\ dx$ and

$$\int \sin(u)\ du = \cos(u) + C, \qquad \int \cos(u)\ du = -\sin(u) + C.$$

The rules for differentiating the other basic trigonometric functions—cosecant, secant, tangent, and cotangent—can be transformed into corresponding integration rules in a similar fashion. Recall the following differentiation rules from HW6.16:

$$\frac{d}{dx}(\csc(f(x))) = -\csc(f(x))\cot(f(x))f'(x),$$

$$\frac{d}{dx}(\sec(f(x))) = \sec(f(x))\tan(f(x))f'(x),$$

$$\frac{d}{dx}(\tan(f(x))) = \sec^2(f(x))f'(x),$$

$$\frac{d}{dx}(\cot(f(x))) = -\csc^2(f(x))f'(x).$$

These differentiation rules can be transformed into the following integration rules:

$$\int \csc(f(x))\cot(f(x))f'(x)\ dx = -\csc(f(x)) + C,$$

$$\int \sec(f(x))\tan(f(x))f'(x)\ dx = \sec(f(x)) + C,$$

$$\int \sec^2(f(x))f'(x)\ dx = \tan(f(x)) + C,$$

$$\int \csc^2(f(x))f'(x)\ dx = -\cot(f(x)) + C;$$

or if $u = f(x)$ and $du = f'(x)\,dx$,

$$\int \csc(u)\cot(u)\ du = -\csc(u) + C, \quad \int \sec(u)\tan(u)\ du = \sec(u) + C,$$

$$\int \sec^2(u)\ du = \tan(u) + C, \qquad \int \csc^2(u)\ du = -\cot(u) + C.$$

Find the following antiderivatives, using substitution and the four new integration rules:

**a.** $\int 3\sec^2(5x)\ dx$

**b.** $\int 2\sec(4x)\tan(4x)\ dx$

**c.** $\int 2x^2\csc^2(x^3)\ dx$

**d.** $\int \dfrac{\csc(\sqrt{r})\cot(r)}{\sqrt{r}}\ dr$

**e.** $\int t\csc^2(t^2+1)\ dt$

**f.** $\int (q + \sec^2(\tfrac{1}{2}q))\ dq$

**g.** $\int \dfrac{1}{\theta^2}\sec\left(\dfrac{1}{\theta}\right)\tan\left(\dfrac{1}{\theta}\right)\ d\theta$

**h.** $\int \sec(7p)(\sec(7p) - \tan(7p))\ dp$

2. Reversing the rules for differentiating the basic trigonometric functions enables you to find the antiderivatives of $\sin(u)$ and $\cos(u)$ and it enables you to find the antiderivatives of the expressions $\sec^2(u)$, $\csc^2(u)$, $\sec(u)\tan(u)$, and $\csc(u)\cot(u)$, but what if you wanted to find the antiderivatives of the other basic trigonometric functions tangent, cosecant,

secant, and cotangent? In parts a and b below, you will use substitution to find rules for integrating $\tan(x)$, $\csc(x)$, $\sec(x)$, and $\cot(x)$ or, more generally, of $\tan(u)$, $\csc(u)$, $\sec(u)$, and $\cot(u)$, and then you will use the new rules to evaluate some antiderivatives.

**a.** You know the rules for finding the antiderivative of sine and cosine:

$$\int \sin(x)\ dx = -\cos(x) + C, \qquad \int \cos(x)\ dx = \sin(x) + C.$$

Use substitution to show that the following integration rules hold for tangent, cotangent, secant, and cosecant:

**(1)** Show that $\displaystyle \int \tan(x)\ dx = -\ln\left|\cos(x)\right| + C,$ provided $\cos(x) \neq 0$.

*Hint: Note that $\int \tan(x)\ dx = \displaystyle\int \frac{\sin(x)}{\cos(x)}\ dx.$ Let $u = \cos(x)$.*

**(2)** Show that $\displaystyle \int \cot(x)\ dx = \ln\left|\sin(x)\right| + C,$ provided $\sin(x) \neq 0$.

*Hint: Note that $\int \cot(x)\ dx = \displaystyle\int \frac{\cos(x)}{\sin(x)}\ dx.$ Let $u = \sin(x)$.*

**(3)** Show that $\displaystyle \int \sec(x)\ dx = \ln\left|\sec(x) + \tan(x)\right| + C.$

*Hint: Note that*

$$\int \sec(x)\ dx = \int \sec(x)\frac{\sec(x) + \tan(x)}{\sec(x) + \tan(x)}\ dx = \int \frac{\sec^2(x) + \sec(x)\tan(x)}{\sec(x) + \tan(x)}\ dx.$$

*Let $u = \sec(x) + \tan(x)$ and recall that $\dfrac{d}{dx}(\sec(x)) = \sec(x)\tan(x)$ and*

$\dfrac{d}{dx}(\tan(x)) = \sec^2(x)$ *(see part 1).*

**(4)** Show that $\displaystyle \int \csc(x)\ dx = \ln\left|\csc(x) - \cot(x)\right| + C.$

*Hint: Note that*

$$\int \csc(x)\ dx = \int \csc(x)\frac{\csc(x) - \cot(x)}{\csc(x) - \cot(x)}\ dx = \int \frac{\csc^2(x) - \csc(x)\cot(x)}{\csc(x) - \cot(x)}\ dx.$$

*Let $u = \csc(x) - \cot(x)$ and recall that $\dfrac{d}{dx}(\csc(x)) = -\csc(x)\cot(x)$ and*
$\dfrac{d}{dx}(\cot(x)) = -\csc^2(x)$ *(see part 1).*

**b.** More generally, if $u = f(x)$, then $du = f'(x)\ dx$ and you know that

$$\int \sin(u)\ du = -\cos(u) + C, \qquad \int \cos(u)\ du = \sin(u) + C.$$

Similarly, it can be shown that

$$\int \tan(u)\ du = -\ln|\cos(u)| + C,$$

$$\int \cot(u)\ du = \ln|\sin(u)| + C,$$

$$\int \sec(u)\ du = \ln|\sec(u) + \tan(u)| + C,$$

$$\int \csc(u)\ du = \ln|\csc(u) - \cot(u)| + C.$$

Evaluate the following indefinite integrals, using the method of substitution in conjunction with these new rules:

**(1)** $\displaystyle\int 2\tan\left(\frac{x}{2}\right) dx$

**(2)** $\displaystyle\int t\sec(t^2)\ dt$

**(3)** $\displaystyle\int \csc(e^r) e^r\ dr$

**(4)** $\displaystyle\int \frac{\cot(\sqrt[3]{z})}{\sqrt[3]{z^2}}\ dz$

**(5)** $\displaystyle\int \frac{1}{\cos(2x)}\ dx$

Hint: $\sec(\theta) = \dfrac{1}{\cos(\theta)}$

**(6)** $\displaystyle\int \frac{2}{\cot(2x)}\ dx$

Hint: $\cot(\theta) = \dfrac{1}{\tan(\theta)}$

**(7)** $\displaystyle\int \tan\frac{(e^{-x})}{e^x}\ dx$

**(8)** $\displaystyle\int \cot(\sin(x))\cos(x)\ dx$

---

## SECTION 2

## Using Integration by Parts

Each rule for integration corresponds to a rule for differentiation, because integration and differentiation are inverse processes. In the last section, you developed integration by substitution, which corresponds to the Chain Rule. In this section, you will develop the method of *integration by parts*, which corresponds to the Product Rule.

Similar to the method of substitution, the idea underlying integration by parts is to replace an integral that you do not know how to evaluate with an expression containing an integral that you can handle. For example, consider the integral

$$\int x\cos(x)\ dx.$$

None of the methods you have learned to date applies in this case. You cannot manipulate $x\cos(x)$ to get a function whose integral you know, and trying to use substitution gets you nowhere. However, using integration by parts, you can show that

$$\int x \cos(x) \ dx = x \sin(x) - \int \sin(x) \ dx$$

and now you can find a solution, since you know how to find the antiderivative of $\sin(x)$.

But, how do you know that evaluating $\int x \cos(x) \ dx$ is equivalent to evaluating $x \sin(x) - \int \sin(x) \ dx$? In particular, what is the formula for integration by parts? How is it related to the Product Rule? How do you use it? First, we will find the general formula for integration by parts, and then, in the next task, we will guide you through using the approach to evaluate $\int x \cos(x) \ dx$.

Suppose $f$ and $g$ are differentiable functions. Then, according to the Product Rule,

$$\frac{d}{dx}(f(x)g(x)) = f(x)g'(x) + g(x)f'(x),$$

or rearranging terms,

$$f(x)g'(x) = \frac{d}{dx}(f(x)g(x)) - g(x)f'(x).$$

If you take the antiderivative of both sides,

$$\int f(x)g'(x) \ dx = \int \left( \frac{d}{dx}(f(x)g(x)) - g(x)f'(x) \right) dx,$$

replace the integral of the difference on the right side by the difference of two integrals,

$$\int f(x)g'(x) \ dx = \int \frac{d}{dx}(f(x)g(x)) \ dx - \int g(x)f'(x) \ dx,$$

and use the fact that integration and differentiation are inverses processes, you get the formula for *integration by parts.*

$$\int f(x)g'(x) \ dx = f(x)g(x) - \int g(x)f'(x) \ dx$$

This looks like a pretty complicated formula. An easier way to write it is to let

$$u = f(x) \qquad \text{and} \qquad dv = g'(x) \ dx;$$

then

$$du = f'(x) \ dx \qquad \text{and} \qquad v = \int g'(x) \ dx = g(x).$$

and substituting gives you

$$\int \underbrace{f(x)}_{u}\underbrace{g'(x) \ dx}_{dv} = \underbrace{f(x)}_{u} \cdot \underbrace{g(x)}_{v} - \int \underbrace{g(x)}_{v}\underbrace{f'(x) \ dx}_{du}.$$

Hence, an alternate formula—in fact, the usual formula—for integration by parts is

$$\int u \, dv = uv - \int v \, du.$$

The next task guides you through how to use integration by parts to transform the integral

$$\int x \cos(x) \, dx$$

into the equivalent expression

$$\int x \cos(x) \, dx = x \sin(x) - \int \sin(x) \, dx,$$

which you *can* then evaluate.

### Task 8-6: Examining the Strategy Underlying Integration by Parts

1. Explain why you cannot evaluate $\int x \cos(x) \, dx$ directly or by using substitution. (Try it and see what happens.)

2. Use the integration by parts formula

$$\int u \, dv = uv - \int v \, du$$

to evaluate

$$\int x \cos(x) \, dx.$$

**a.** The first thing you need to do is to match the left side of the integration by parts formula $u \, dv$ with $x \cos(x) \, dx$, by choosing $u = f(x)$ and $dv = g'(x) \, dx$, where $dv$ includes the $dx$ part. In the case of $x \cos(x) \, dx$, there are two possibilities. You could let

$$u = x \quad \text{and} \quad dv = \cos(x) \, dx,$$

or since $x \cos(x) \, dx = \cos(x) x \, dx$, you could let

$$u = \cos(x) \quad \text{and} \quad dv = x \, dx.$$

It turns out that the first choice for $u$ and $dv$ works, whereas the second one does not (you will have an opportunity to reflect on why this is the case in part 3). For now, assume that $u = x$ and $dv = \cos(x)\ dx$.

Before you can apply the formula for integration by parts, you need to find $du$ and $v$ using your choices for $u$ and $dv$.

(1) In general, $u = f(x)$, and as a result, $du = f'(x)\ dx$. Use differentiation to find $du$, when $u = x$.

$$u = x,$$

$$du = \underline{\hspace{2cm}}.$$

(2) In general, $dv = g'(x)\ dx$. Consequently, $v = \int dv = \int g'(x)\ dx = g(x)$. Use integration to find $v$, when $dv = \cos(x)\ dx$.

$$dv = \cos(x)\ dx,$$

$$v = \int \cos(x)\ dx = \underline{\hspace{2cm}}.$$

(3) Summarize your results.

$$u = \underline{\hspace{1.5cm}}, \qquad dv = \underline{\hspace{1.5cm}},$$

$$du = \underline{\hspace{1.5cm}}, \qquad v = \underline{\hspace{1.5cm}}.$$

**b.** Now that you have $u$, $du$, $v$, and $dv$, you are ready to apply the integration by parts formula.

(1) Substitute into the formula.

$$\int u\ dv = uv - \int v\ du$$

$$\int \underbrace{x}_{u}\ \underbrace{\cos(x)\ dx}_{dv} = \underline{\hspace{5cm}}.$$

(2) You should have $\int x \cos(x)\ dx = x \sin(x) - \int \sin(x)\ dx$. If you do not, retrace your steps and find where you went astray.

**c.** If the new integral on the right side is one you know how to solve, then you have made appropriate choices for $u$ and $dv$ and integration by parts works. This is what happens in this case. Find $\int x \cos(x)\ dx$ by evaluating the right side of the equation.

$$\int x \cos(x)\ dx = x \sin(x) - \int \sin(x)\ dx,$$

$$\int x \cos(x)\ dx = \underline{\hspace{5cm}}.$$

**d.** Check your solution by showing that the derivative of your result equals $x \cos(x)$. Observe that you need to use the Product Rule to do this.

**3.** Examine what happens if you make the other choice for $u$ and $dv$.

In part 2, we suggested that you choose

$$u = x \quad \text{and} \quad dv = \cos(x)\ dx$$

and integration by parts worked, since you were able to evaluate the new integral on the right side of the equation. Suppose, instead, you had chosen

$$u = \cos(x) \quad \text{and} \quad dv = x\ dx.$$

Show why this choice does not work.

**a.** Find $du$ and $v$ in this case.

**(1)** Assume $u = \cos(x)$. Use differentiation to find $du$.

**(2)** Assume $dv = x\ dx$. Use integration to find $v$.

**b.** Substitute into the integration by parts formula, using the alternate expressions for $u$, $du$, $v$, and $dv$.

$$\int u\ dv = uv - \int v\ du,$$

$$\int \underbrace{x}_{u}\ \underbrace{\cos(x)\ dx}_{dv} = \underline{\hspace{4cm}}.$$

**c.** Explain why integration by parts does not work with this choice of $u$ and $dv$.

As with substitution, the general idea underlying integration by parts is to transform an integral that you do not know how to solve into an expression containing an integral that you can solve. For example, in Task 8-6, you did not know how to integrate $x\cos(x)$, but you could integrate $\sin(x)$ and, consequently, integration by parts worked in this case.

In general, the method of integration by parts involves doing the following steps:

• Match the given integral to $\int u\ dv$. Choose $u$ and $dv$.

*Note: dv contains the dx part.*

- Find *du* by differentiating *u* and find *v* by integrating *dv*.

    *Note: You may have to use the method of substitution, or possibly even integration by parts (again), to find v.*

- Substitute into the integration by parts formula, using the expressions for *u*, *du*, *v*, and *dv*:

$$\int u \, dv = uv - \int v \, du.$$

- Evaluate the right side of the equation.

    *Note: If you are unable to evaluate the integral on the right, try another choice for u and dv. If no other choice works, then integration by parts does not apply.*

- Check the result using differentiation.

    *Note: You will have to use the Product Rule.*

In the next task, you will use integration by parts to evaluate some integrals. As with substitution, it may take more than one try to get the answer. Persistence helps. So does patience.

---

## Task 8-7: Using Integration by Parts

1. Use integration by parts to find the following antiderivatives.

    **i.** For each problem, clearly indicate your choices for *u* and *dv* and show how you find *du* and *v*.

    **ii.** Check your answer by differentiation.

    **a.** $\int x \sin(x) \, dx$

**b.** $\int xe^x\, dx$

**c.** $\int x\ln(x)\, dx$

**d.** $\int (x+8)^{10}(x+1)\, dx$

**2.** Use integration by parts to evaluate the following definite integrals.

**a.** $\int_0^{1/3} xe^{3x}\, dx$

**b.** $\displaystyle\int_0^3 x\sqrt{x+1}\ dx$

In the last task of this section, you will tackle a project where you need to use integration by parts. The question you wish to answer is: How much energy does it take to sound off?

Suppose Kool Klassical Kitty is exercising her vocal chords by singing a pure note. The sounds that emerge result from the oscillation of her vocal chords. The sound pressure, when she is singing at a steady sound level, can be described quite nicely by a sinusoidal function as shown in Figure 1.

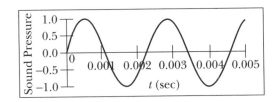

Figure 1. Sound pressure, $w(t) = \sin(2\pi f t)$ Newtons/meter$^2$.

Many singers "attack" notes by increasing the volume of air forced past the vocal chords by the chest cavity as time goes by. For example, it is possible for Kool Kitty to force air past her vocal chords so that the pressure amplitude of the sound—that is, the maximum possible value of the pressure at a given time—increases as the square root of time as shown in Figure 2.

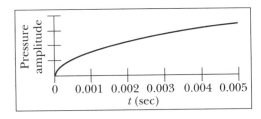

Figure 2. Pressure amplitude, $g(t) = c\sqrt{t}$.

This means that the sound pressure emerging from her mouth is the product of two functions—a sine function which represents the variation of sound pressure and a square root function which determines the pressure amplitude of the sound.

The energy given to the molecules of air in the room when sound is transmitted is the sound energy. This sound energy can be calculated by integrating a function related to the product of the function representing the attack (Fig. 2) and the one describing the oscillation of Kitty's vocal chords (Fig. 1). When the attack function is a square root function, the integral representing the total sound energy between times $a$ and $b$ has the form $\int_a^b t\sin^2(t)\,dt$, which can be evaluated using a trigonometric substitution and integration by parts.

---

## Task 8-8: (Project) Sounding Off

1. Examine Kool Kitty's waveform.

If a pure tone is voiced with a constant pressure amplitude, then the *shape* of the sound wave is sinusoidal. Assume the waveform when Kool Kitty sings a concert A is given by the function

$$w(t) = \sin(2\pi f t) \text{ Newtons/meter}^2,$$

where $f$ is the frequency of the sound in cycles per second and $t$ is the time in seconds.

  **a.** Examine the shape of the waveform when Kool Kitty sings a concert A at 440 cycles per second—that is, when $f = 440$ Hz. Use your CAS to graph her waveform for the short time between 0 and 0.01 seconds and place it in the space below. (This short time interval will enable you to see the shape of $w$ more clearly.)

  **b.** Since one-hundredth of a second (0.01 second) is not much time, assume that the musical kitty actually sings a quick staccato note for a longer period of time.

  **(1)** Use your CAS to examine the shape of the graph of $w$ for some longer time intervals. In particular, compare the graphs of $w$, where $t$ varies from 0 to 0.02, from 0 to 0.03, from 0 to 0.04, from 0 to 0.05, and from 0 to 0.1 seconds.

(2) Assume Kool Kitty sings the note for one-tenth of a second. Print a copy of the graph of her waveform $w$ for this longer time interval. Place your graph in the space below.

2. Examine Kool Kitty's pressure amplitude.

Suppose Kool Kitty keeps pushing harder with her diaphragm as the brief tenth of a second goes by, so that the attack on her note has a pressure amplitude as a function of time given by

$$g(t) = 9.0\sqrt{t}.$$

Graph the pressure amplitude $g$ in the time interval between 0 and 0.1 seconds. Place a copy of your graph in the space below.

3. Find an equation for Kool Kitty's sound pressure.

In general, the sound pressure $h$ varies in time as the product of the pressure amplitude $g$ and the waveform $w$.

a. Represent Kool Kitty's sound pressure $h$ by an expression by multiplying the expression representing her pressure amplitude $g(t)$ and the expression representing her waveform $w(t)$.

$$h(t) = \underline{\hspace{3cm}}.$$

b. Graph Kool Kitty's sound pressure $h$ in the time interval between 0 and 0.1 seconds. Place a copy of your graph in the space below.

**4.** Find an equation for Kool Kitty's sound intensity.

Physicists are often interested in the emitted energy passing through a unit of area due to a passing sound wave. The energy is really a measure of the jiggling of air molecules as the sound wave passes through the air. The rate at which sound energy is emitted per unit area is known as the sound intensity. In air at room temperature, the sound intensity is related to the sound pressure by the expression

$$I(t) = \left(\frac{1}{28.7}\right)^2 h(t)^2 \text{ meter}^2/\text{Watts}.$$

Use your expression for $h$ from part 3 to show that the sound intensity for Kool Kitty as a function of time is given by

$$I(t) = ct\sin^2(kt) \text{ meter}^2/\text{Watts},$$

where $c = 9.8 \times 10^{-2}$ and $k = 2.76 \times 10^3$.

**5.** Calculate Kool Kitty's sound energy.

If the sound intensity were constant during a given time interval, then the sound energy could be found by using the formula

$$\text{Energy} = \text{intensity} \cdot \text{time}.$$

In Kool Classical Kitty's case, the intensity is varying, so the energy per unit area from 0 seconds to $x$ seconds is given by

$$E(x) = \int_0^x I(t)\ dt.$$

In order to find the energy per unit area in Kitty's note, you need to integrate Kitty's sound intensity from 0 seconds to 0.1 seconds—that is, you need to find

$$E(0.1) = \int_0^{0.1} I(t)\ dt = c\int_0^{0.1} t\sin^2(kt)\ dt \text{ meter}^2/\text{Joules}.$$

Unfortunately you do not know how to evaluate this integral directly. You can, however, replace it with an equivalent integral by using a trigonometric identity, and then evaluate the new integral using integration by parts. We will guide you through this process. As you do the steps, you should begin to appreciate what scientists and social scientists have to go through to solve some of the integrals that interest them.

**a.** First, find an equivalent expression for $c \int_0^{0.1} t \sin^2(kt) \, dt$. In particular, show that

$$c \int_0^{0.1} t \sin^2(kt) \, dt = \frac{c}{2} \left( \int_0^{0.1} t \, dt - \int_0^{0.1} t \cos(2kt) \, dt \right).$$

This is accomplished by doing the following:

**i.** Replace $\sin^2(kt)$ using the trigonometric identity

$$\sin^2(\theta) = \frac{1}{2} - \frac{1}{2} \cos(2\theta).$$

**ii.** Manipulate the expression using the facts that the integral of the sum of two functions equals the sum of the integrals, and the integral of a constant times a function equals the constant times the integral of the function.

$$c \int_0^{0.1} t \sin^2(kt) \, dt =$$

**b.** Evaluate the first integral on the right-hand side of the equation in part a directly.

$$\int t \, dt =$$

**c.** Keeping in mind that $k$ is a constant, use integration by parts to evaluate the second integral on the right-hand side of the equation in part a.

$$\int t \cos(2kt) \, dt =$$

**d.** Show that the energy per unit area associated with Kitty's staccato note is given by

$$E(0.1) = c \int_0^{0.1} t \sin^2(kt) \, dt = \frac{c}{2}\left( \frac{t^2}{2} - \frac{1}{2k} t \sin(2kt) - \frac{1}{4k^2} \cos(2kt) \right)\Bigg|_{t=0}^{t=0.1}$$

by combining your results from parts a through c.

**e.** Now for the number! Find $E(0.1)$, the total sound energy in Joules/meter$^2$ associated with Kool Klassical Kitty's staccato concert A note that only lasts a tenth of a second, by evaluating the right side of the equation in part d, where

$$c = 9.8 \times 10^{-2} \quad \text{and} \quad k = 2.76 \times 10^3.$$

Do not be surprised when this turns out to be a small number.

**f.** Check your result by using your CAS to evaluate

$$c \int_0^{0.1} t \sin^2(kt) \, dt, \quad \text{where } c = 9.8 \times 10^{-2} \text{ and } k = 2.76 \times 10^3.$$

## Summary of Basic Rules of Integration

The following list contains all the integration rules which you have encountered (or in the case of $\int \ln(u) \, du$, which you will encounter) in various tasks and homework exercises:

**1.** $\int u \, dv = uv - \int v \, du$

**2.** $\int u^n \, du = \frac{1}{n+1} u^{n+1} + C, \quad n \neq -1$

**3.** $\int \frac{1}{u} \, du = \ln|u| + C$

**4.** $\int e^u \, du = e^u + C$

**5.** $\int a^u \, du = \frac{1}{\ln a} a^u + C$

**6.** $\int \ln(u) \ du = u \ln(u) - u + C$

**7.** $\int \sin(u) \ du = -\cos(u) + C$

**8.** $\int \cos(u) \ du = \sin(u) + C$

**9.** $\int \sec^2(u) \ du = \tan(u) + C$

**10.** $\int \csc^2(u) \ du = -\cot(u) + C$

**11.** $\int \sec(u) \tan(u) \ du = \sec(u) + C$

**12.** $\int \csc(u) \cot(u) \ du = -\csc(u) + C$

**13.** $\int \tan(u) \ du = -\ln|\cos(u)| + C$

**14.** $\int \cot(u) \ du = \ln|\sin(u)| + C$

**15.** $\int \sec(u) \ du = \ln|\sec(u) + \tan(u)| + C$

**16.** $\int \csc(u) \ du = \ln|\csc(u) - \cot(u)| + C$

## Unit 8 Homework After Section 2

- Complete the tasks in Section 2. Be prepared to discuss them in class.

- Use integration by parts to find some antiderivatives and solve some definite integrals in HW8.6.

**HW8.6** Evaluate the following integrals using integration by parts. Check the antiderivatives by differentiating

**1.** $\int x^2 \ln(x) \ dx$

**2.** $\int \frac{\ln(x)}{\sqrt{x}} \ dx$

**3.** $\int x e^{-2x} \ dx$

**4.** $\int x^{-3} \ln(3x) \ dx$

**5.** $\int (x+1) \ln(x) \ dx$

**6.** $\int_{\pi}^{3\pi/2} 3x \cos(x) \ dx$

**7.** $\int_{5}^{12} x\sqrt{x+4} \ dx$

**8.** $\int x \sin(4x) \ dx$

**9.** $\int x(2x+1)^{99} \ dx$

**10.** $\int x^3 \sqrt{x^2+1} \ dx$

**∞**

- Use integration by parts to analyze some "real" situations in HW8.7.

**HW8.7** Analyze the following situations.*

1. The rate of electrical energy used by a family, in kilo-watt hours per day, is given by

$$K(t) = 10te^{-t},$$

where $t$ is the time in hours and $0 \le t \le 24$.

   a. How many kilowatt hours does the family use in the first $T$ hours of a day?

   b. Use your result from part a to determine the number of kilowatt hours the family uses in the first 4 hours of a day.

2. Suppose an oral dose of a drug is taken. From the time the drug is taken, the drug is assimilated in the body and excreted through the urine. The total amount of the drug that has passed through the body at time $T$ is given by

$$\int_0^T E(t)\ dt.$$

A typical rate of excretion function is

$$E(t) = te^{-kt},$$

where $k > 0$ is a constant and $t$ is time in hours.

   a. Find a formula for

$$\int_0^T E(t)\ dt.$$

   b. Use your result from part a to determine the amount of the drug that has passed through the body in 10 hours, where $k = 0.2$ milligram/hour.

- Use integration by parts to find the antiderivative of $\ln(x)$ in HW8.8.

**HW8.8** Use integration by parts to show that

$$\int \ln(x)\ dx = x\ln(x) - x + C.$$

- Use integration by parts several times to evaluate the integrals in HW8.9.

**HW8.9** Use integration by parts to evaluate the following integrals. In each exercise, you will have to use integration by parts more than one time.

1. $\int x^2 e^x\ dx$

2. $\int x^2 \sin(4x)\ dx$

3. $\int x^3 \sqrt{x+1}\ dx$

4. $\int (\ln(x))^2\ dx$

---

*See *Applied Calculus* by Bittenger and Morrell, Addison-Wesley, New York, 1993.

## SECTION 3

### Using Integration Tables

The list at the end of the last section contains all the integration rules which you have encountered in the tasks and homework exercises. The list is far from complete. Years ago, a favorite pastime of mathematicians was trying to develop techniques to solve various classes of integrals. They compiled their results in *integration tables*, which give the antiderivatives for functions which satisfy general forms. Prior to the advent of computer packages that do symbolic manipulations, integration tables were an important tool for scientists who need to solve complex integrals. The Appendix in the back of the book contains a typical table. To use the table, you need to do the following:

- Find the section in the table that lists the integrals containing the same basic expression as the integral you wish to solve, where the basic expressions have forms such as

$$ax + b,$$
$$p^2 - x^2,$$
$$ax^2 + bx + c,$$
$$\sin(ax).$$

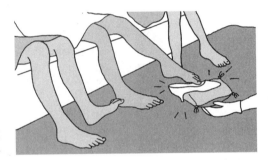

- Match the given integral with one that appears in that section of the table.
- Find the solution for the given integral by plugging in the appropriate values for the constants that appear in the solution for the general form specified in the table.

For example, to use the table in the Appendix to solve

$$\int \frac{dx}{x(3 + x)},$$

you first need to recognize that this integral contains the basic expression

$$ax + b,$$

where $a = 1$ and $b = 3$. If you look through the list of integrals in the section for $ax + b$, you will see that the given integral has the same form as Equation (55):

$$\int \frac{dx}{x(ax + b)},$$

whose solution is

$$\frac{1}{b} \log \left| \frac{x}{ax + b} \right| + C,$$

where log is the natural logarithm. Consequently, you can find the solution to the given integral by substituting $a = 1$ and $b = 3$ in the solution for the general form:

$$\int \frac{dx}{x(3 + x)} = \frac{1}{3} \ln \left| \frac{x}{x + 3} \right| + C.$$

To be able to use the tables, you need to become familiar with the way the tables are organized, and you need to become adept at recognizing the general form of the integrals you wish to evaluate. Task 8-9 will give you some practice doing this.

## Task 8-9: Using the Tables

1. Familiarize yourself with the headings for the different sections in the table contained in the Appendix.

   Flip through the pages in the table and note the heading of each section. For instance, the first section lists "Expressions Containing $ax + b$"; this is the section where you would look up integrals such as

   $$\int \frac{x}{(6x - 2)^3} \, dx \quad \text{or} \quad \int t^{-4}(2 + 6t)^{10} \, dt,$$

   since $(6x - 2)$ and $(2 + 6t)$ have the general form $ax + b$.

2. Use the tables to evaluate the following integrals. In each case, given the number(s) of the formula(s) in the table which you used. Note that, for a given problem, it is possible that more than one formula is applicable. Also note that you may have to use the same formula several times or several different formulas to arrive at a final solution (which does not contain any integrals).

   a. $\int x(4x^2 - 9)^3 \, dx$

   b. $\int \sin^3(2r) \, dr$

c. $\int \dfrac{dm}{m(m^2 + 5m + 4)}$

d. $\displaystyle\int_0^6 \sqrt{36 - x^2}\ dx$

e. $\displaystyle\int e^{-3x} \cos(x)\ dx$

f. $\displaystyle\int_0^{\pi/2} \sin^2(\theta)\ \cos^4(\theta)\ d\theta$

Use the integration tables to do a project which involves trying to answer the question: How high should the water in Martha's fish bowl be?

Enrique's pet goldfish Martha has been unhappy because she does not have enough space to play. A fish doctor has suggested that a healthy goldfish should have at least 4 liters of water. Enrique does not have any way to measure the volume of water in liters. Fortunately, Martha lives in a special fish bowl which has a uniform thickness and where each side is shaped like a circle with a rectangular wedge in the middle, and the bowl manufacturer has provided specifications for the bowl with all the dimensions listed.

Enrique has a ruler with a centimeter (cm) scale and he knows that you are taking calculus. He would like you to help him figure out how high the water should be in order to provide Martha with 4 liters of water. To do this, you need to integrate the function that describes the curvature of the bowl. Solving the resulting integral by hand involves using some sophisticated techniques of integration, which you have not studied. You can however, use your knowledge of calculus and the table lookup method to help Enrique calculate the height to which he should fill Martha's fish bowl so it contains 4 liters of water.

In order to calculate the volume of water the bowl holds, you need to know more about the size and shape of the bowl. The shape of the front surface of Martha's bowl consists of two semicircles of radius $R$ separated by a rectangle of width $w$ and height $2R$. This is shown in the diagram below.

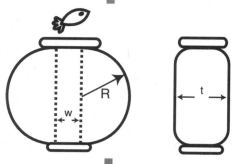

Radius: $R = 12$ cm

Wedge: $w = 6$ cm

Width: $t = 15$ cm

Volume of 4 liters = 4000 cm$^3$

Using calculus to help Enrique figure out the height he needs to fill his fish bowl in order to provide the poor fish Martha with 4 liters, or 4000 cm$^3$, of water is not easy. Why doesn't Enrique have something to measure water volumes in cubic centimeters?! In Task 8-10, you will help Enrique solve this problem. The strategy requires several steps. You will need to do the following:

• Find the function that describes the curvature of Martha's bowl.
• Find the function that gives the area of the face of her bowl when the water level is at height $h$, where $0 \le h \le 2R$, by using the table lookup method to integrate the curvature function.
• Find the volume of the water in the bowl as a function of $h$ by multiplying the area function and the width of the bowl.
• Find the value of $h$ so that the volume of water in Martha's bowl is exactly 4000 cubic centimeters (cm$^3$).

## Task 8-10: (Project) Finding the Right Water Level

**1.** Model the situation.

In order to turn this into a calculus problem, put the bowl on its side (forget about the water for now), and position it on a pair of axes, with the origin at the center of the bowl. This will help you figure out how to determine the function describing the curvature of Enrique's bowl. The tipped bowl diagram is shown below.

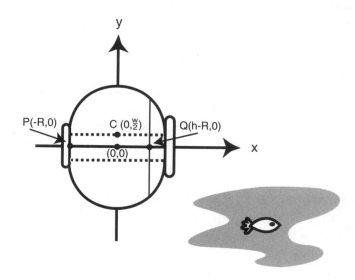

**a.** Study the diagram in the figure carefully. Explain why it makes sense to label the three points $P(-R,0)$, $C(0, w/2)$, and $Q(h - R, 0)$.

**b.** Show that the following function describes the curvature of the bowl:

$$y = \sqrt{R^2 - x^2} + \frac{w}{2}.$$

*Hint: In general, the equation for a circle with center $C(a, b)$ and radius $r$ is $(x - a)^2 + (y - b)^2 = r^2$. The center of the circular portion of Martha's bowl is $C(0, w/2)$ and the radius is $R$.*

**c.** In Martha's case, $R = 12$ and $w = 6$. Find the equation for $y$ with these dimensions.

**2.** Let $A(h)$ be the surface area of the bowl which is covered by water when the height of the water is $h$, where $0 \leq h \leq 24$. Represent the function $A$ by an expression.

**a.** Explain why the following definite integral represents half of the surface area covered by the water.

$$\frac{A(h)}{2} = \int_{-12}^{h-12} (\sqrt{144 - x^2} + 3) \, dx, \quad \text{for } 0 \leq h \leq 24.$$

**b.** Represent $A$ by an expression by evaluating the integral

$$A(h) = 2 \int_{-12}^{h-12} (\sqrt{144 - x^2} + 3) \, dx.$$

**(1)** Use the properties of integrals to show that

$$A(h) = 2 \int_{-12}^{h-12} (\sqrt{144 - x^2}) \, dx + 2 \int_{-12}^{h-12} 3 \, dx.$$

**(2)** Evaluate the first integral in part (1) using the table lookup method. Your result will be in terms of $h$.

$$2 \int_{-12}^{h-12} \left( \sqrt{144 - x^2} \right) dx =$$

**(3)** Evaluate the second integral in part (1) directly. Again, your result will be in terms of $h$.

$$2 \int_{-12}^{h-12} 3 \, dx =$$

**(4)** Find an expression for $A(h)$, when $0 \le h \le 24$, by combining your results from parts (1) through (3).

3. Let $V(h)$ be the volume of the water in the bowl when the height of the water is $h$.

   **a.** You can find $V(h)$ by multiplying the surface area and the width of the bowl, which, in this case, is 15 cm; that is,

   $$V(h) = 15A(h).$$

   Explain why this is a reasonable thing to do.

   **b.** Plot $V(h)$ for $0 \le h \le 24$, using your CAS. Place a copy of your graph in the space below.

   **c.** Use your CAS to find the volume of the water when

      **(1)** $h = 6$ cm

      **(2)** $h = 18$ cm

**4.** Analyze the situation.

   **a.** Use your graph of $V$ to estimate the height in centimeters when the volume of the water in the bowl is 4000 cm³.

   **b.** Use your CAS to find the height in centimeters when the volume of the water in the bowl is 4000 cm³. Finally, Enrique can use his ruler, which has a centimeter scale, to fill the fish bowl to exactly the right level and keep Martha happy!

## Unit 8 Homework After Section 3

- Complete the tasks in Section 3. Be prepared to discuss them in class.

- Use table lookup to find some antiderivatives and evaluate some definite integrals in HW8.10.

**HW8.10** Evaluate the following integrals using the table lookup method.

> *Notes: You may have to pull a constant outside the integral. You may have to use the tables more than once on a given problem.*

**1.** $\displaystyle \int \frac{2}{5x(7x+2)}\, dx$        **4.** $\displaystyle \int t^4 e^{5t}\, dt$

**2.** $\displaystyle \int \frac{-5\ln(x)}{x^3}\, dx$        **5.** $\displaystyle \int_0^\pi \sin^4\!\left(\frac{\theta}{2}\right)\cos\!\left(\frac{\theta}{2}\right) d\theta$

**3.** $\displaystyle \int \sqrt{4m^2 + 16}\, dm$        **6.** $\displaystyle \int_3^5 \frac{dx}{x^3 - 4x^2 + 4x}$

- Review the methods of integration that you have learned in HW8.11.

**HW8.11**

**1.** Use the method of algebraic manipulation, substitution, or integration by parts to evaluate the following integrals. State which method you are using. Show all your work. Check your answer by differentiating.

   **a.** $\displaystyle \int \left(6x^2 + \frac{3}{x^3} - 15\right) dx$        **c.** $\displaystyle 7\int e^t\, dt$

   **b.** $\displaystyle \int_{-2}^0 (t^3 - 3t^2 + 3)^3 (t^2 - 2t)\, dt$        **d.** $\displaystyle \int x^{-4}\cos(x^{-3})\, dx$

**e.** $\int \dfrac{2.5}{r}\,dr$

**k.** $\int_{-1}^{1} 6t\sqrt[5]{9-t}\,dt$

**f.** $\int_{1}^{2} \dfrac{2r^3 - r + 8}{r^4 - r^2 + 16r}\,dr$

**l.** $\int \ln(6x)\,dx$

**g.** $\int (x-1)\,e^{2x^2-4x}\,dx$

**m.** $\int_{0}^{\pi/2} \left(3\cos(\theta) - \tfrac{1}{2}\sin(\theta)\right)\,d\theta$

**h.** $\int_{1}^{3} \left(x^2 - \dfrac{2}{x}\right)^2\,dx$

**n.** $-\int \sin\left(\dfrac{\theta}{\pi}\right)\,d\theta$

**i.** $\int \dfrac{\ln(x)}{x^4}\,dx$

**o.** $\int 2x\sin(x/2)\,dx$

**j.** $\int xe^{-x}\,dx$

2. Use the integration tables to find the following antiderivatives:

**a.** $\int (\ln(3x))^3\,dx$

**d.** $5\int \dfrac{t^2}{(4t^2 - 9)^{3/2}}\,dt$

**b.** $\int \sqrt{\dfrac{x-6}{x+3}}\,dx$

**e.** $\int \cos^4(-2\theta)\,d\theta$

**c.** $\int \dfrac{x^2}{9x^2 + 12x + 4}\,dx$

3. Explain what is occurring here.

   **a.** Give five basic, general cases where you can find the antiderivative directly (without using a "trick," such as substitution). In each case, find the antiderivative.

   **b.** Under what circumstances would you try substitution?

   **c.** Describe the process you use to find an antiderivative using substitution.

   **d.** Describe the process you use to solve a definite integral using substitution.

   **e–g.** Same as parts **b** through **d**, except answer the questions for integration by parts.

## SECTION 4

## Approximating Definite Integrals

What if none of the integration techniques applies and even the integration tables are not helpful? For instance, what if you need to evaluate the definite integral of a function represented by an expression that does not have a simple antiderivative? Or, what if you need to evaluate the definite integral associated with a function represented by a set of data points? In this section, you will examine how to approximate definite integrals in these types of situations by applying the Trapezoidal Rule or by fitting a curve to the data and then integrating the curve-fit function. In addition, you will get some practical experience using integration concepts to explore real-world phenomena in the natural and social sciences as you undertake projects that include estimating the national debt and estimating the average temperature during a given time period.

Let us start by looking at the Trapezoidal Rule. One way to approximate a definite integral is to use *numerical integration*. The general approach is to cover the region bounded by the function with shapes whose areas you can calculate and then add up the areas of the subregions. This is the approach that you used to develop the definition of definite integral. In particular, you covered a given region with rectangles of equal widths and then observed that as the number of rectangles increased, the approximation got closer and closer to the actual value. In the process, you developed a formula for approximating a definite integral using $n$ rectangles, namely

$$\sum_{i=1}^{n} f(x_i)\Delta x,$$

where

$$\Delta x = \frac{b-a}{n} \quad \text{and} \quad x_i = a + i\Delta x, \quad 1 \le i \le n.$$

The rectangular approach gives a fairly accurate estimate, especially when the number of subdivisions is large. Covering the region with trapezoids, however, usually gives a better estimate, and it can be fairly accurate even with a small number of subdivisions.

You used a trapezoidal approach before by simply finding the area of each trapezoid and summing the results. The goal of the next task is to develop a general formula for approximating a definite integral using a trapezoidal approach. This formula is called the *Trapezoidal Rule.*

## Task 8-11: Finding a Formula for the Trapezoidal Rule

Suppose $f$ is a continuous function defined on the closed interval $[a, b]$. Find a formula for using the Trapezoidal Rule to approximate $\int_{a}^{b} f(x) \, dx$.

1. Sketch a graph of $f$. Label the graph and sketch a few trapezoids; that is, on the pair of axes given below:

   **i.** Mark the values of the lower and upper bounds of the interval, $a$ and $b$, on the $x$-axis.

   **ii.** Over $[a, b]$, sketch the graph of $f$ (make $f$ curvaceous and make it lie above the $x$-axis).

**iii.** As usual, let $n$ be the number of trapezoids under consideration, and let

$$\Delta x = \frac{b-a}{n} \quad \text{and} \quad x_i = a + i\Delta x \quad \text{for } 0 \le i \le n.$$

Label $x_0$, $x_1$, $x_2$, $x_3$, $x_{n-1}$, and $x_n$ on the x-axis. Observe that the distance between each successive pair of x's is $\Delta x$, $x_0 = a$ and $x_n = b$.

**iv.** Sketch the trapezoids whose heights are determined by evaluating $f$ at $x_0$ and $x_1$, at $x_1$ and $x_2$, at $x_2$ and $x_3$, and at $x_{n-1}$ and $x_n$.

2. Recall that the area of a trapezoid equals $\frac{1}{2}(h_1 + h_2)b$, where $h_1$ and $h_2$ are the heights of the sides, and $b$ is the width of the base. Find general formulas for the areas of some of the trapezoids. Your formulas should be in terms of $f$, $x_i$, and $\Delta x$.

   **a.** Find a formula for the area of the first trapezoid.

   **b.** Find a formula for the area of the second trapezoid.

   **c.** Find a formula for the area of the third trapezoid.

   **d.** Find a formula for the area of the $n$th trapezoid.

3. Use the trapezoidal approach to find a formula that approximates the value of $\int_a^b f(x)\ dx$ by summing your results from part 2.

**4.** Show that the formula that you developed in part 3 is equivalent to the following formula:

$$\left(\tfrac{1}{2}f(a) + f(x_1) + f(x_2) + f(x_3) + \cdots + f(x_{n-1}) + \tfrac{1}{2}f(b)\right)\Delta x$$

or

$$\left(\tfrac{1}{2}f(a) + \sum_{i=1}^{n-1} f(x_i) + \tfrac{1}{2}f(b)\right)\Delta x.$$

This is called the Trapezoidal Rule.

**5.** Explain why, in general, the Trapezoid Rule gives a better approximation than the rectangular approach, when $n$ and $\Delta x$ are the same size. Support your explanation with a diagram.

In the next three tasks, you will use the Trapezoidal Rule to approximate definite integrals. Doing the calculations by hand is much too tedious. Alternatives are to use your calculator, your CAS, a spreadsheet package, or ISETL. Recall that when you used ISETL to calculate Riemann sums using a rectangular approach (see Task 5-11 or your ISETL handout), the ISETL syntax matched the associated mathematical notation very closely. It is not surprising that the same observation holds when you use ISETL to implement the Trapezoidal Rule. Figure 3 relates the mathematical notation for the Trapezoidal Rule to the associated ISETL syntax. Before continuing on, study the chart carefully, noting the similarities between the two notations.

The Trapezoidal Rule enables you to find a fairly accurate approximation of a definite integral. It is especially useful in cases for which you are unable to solve the integral directly. In particular, the rule can be used to estimate the definite integral of a function which is represented by an expression that does not have a simple antiderivative. This is true, for instance, of functions that have the form $f(x) = be^{ax^2}$. Integrating functions of this type is extremely important in statistics.

For example, suppose the mean or average grade on a test is 75 ($\mu = 75$) and the standard deviation is 10 ($\sigma = 10$) (which means that roughly two-thirds of the scores are between 65 and 85). If the scores are *normally dis-*

| Interpretation | Math notation | ISETL syntax |
|---|---|---|
| function definition | $f(x) = \cdots$ <br><br> or <br><br> $\dfrac{x \mid \cdots}{f(x) \mid \cdots}$ | f := func(x); <br> ... <br> end func; <br> or <br> f := {[..,..],...,[..,..]}; |
| left endpoint of interval | $a = \cdots$ | a := ...; |
| right endpoint of interval | $b = \cdots$ | b := ...; |
| # of trapezoids | $n = \cdots$ | n := ...; |
| width of ith trapezoid | $\Delta x = \dfrac{b - a}{n}$ | delta_x := (b − a)/n; |
| right endpoints of subintervals | $x_i = a + i\Delta x, \quad 1 \le i \le n$ | x := [a + i*delta_x : i in [1..n]]; |
| Trapezoid Rule | $\left(\dfrac{1}{2}f(a) + \displaystyle\sum_{i=1}^{n-1} f(x_i) + \dfrac{1}{2}f(b)\right)\Delta x$ | (0.5*f(a) + %+[f(x(i)) : i in [1..n − 1]] + 0.5*f(b))*delta_x; |

Figure 3.

*tributed*, then you can expect that the number of scores above 75 is about the same as the number of scores below 75. In addition, a person is more likely to score in the mid-70s than in the 50s or the 90s, and the further a possible score is from 75, the less likely it is that someone received that score.

In general, if a random variable is normally distributed with mean $\mu$ and standard deviation $\sigma$, the *probability density function* is given by

$$f(x) = \frac{1}{\sigma\sqrt{2\pi}} e^{-\frac{1}{2}\left(\frac{x-\mu}{\sigma}\right)^2}$$

and the probability that $x$ lies between $a$ and $b$ can be found by integrating the probability density function—that is,

$$P(a \le x \le b) = \int_a^b f(x)\ dx.$$

For instance, returning to the example where $\mu = 75$ and $\sigma = 10$, the probability that a score is between 80 and 90 is given by

$$P(80 \le x \le 90) = \int_{80}^{90} f(x)\ dx = \frac{1}{10\sqrt{2\pi}} \int_{80}^{90} e^{-\frac{1}{2}\left(\frac{x-75}{10}\right)^2}\ dx.$$

Use the Trapezoidal Rule to estimate some probabilities in the next task.

*Note: In HW6.24, you showed that the probability density function has a local maximum at $x = \mu$ and inflection points at $x = \mu + \sigma$.*

## Task 8-12: Using the Trapezoidal Rule on a Function Without a Simple Antiderivative

1. Before estimating some probabilities, examine how the value of the mean $\mu$ and the value of the standard deviation $\sigma$ affect the shape of the graph of a normal distribution function.

   **a.** Use your CAS to graph the probability density function $f$

   $$f(x) = \frac{1}{\sigma\sqrt{2\pi}} e^{-\frac{1}{2}\left(\frac{x-\mu}{\sigma}\right)^2}$$

   with $\mu = 75$ and $\sigma = 10$, for $50 \leq x \leq 100$.

   Print *three* copies of the graph. Place one copy below and set the other two aside.

   **b.** Write a couple of sentences describing the shape of the graph of $f$.

   **c.** At what value of $x$ does the function $f$ appear to have a local maximum?

   **d.** Use your CAS to examine how varying the value of the mean $\mu$ affects the location and shape of the graph of the probability density function.

   Fix the value of the standard deviation $\sigma$ to be 10. Graph the probability density function for several different values of $\mu$. Summarize your observations.

   **e.** Use your CAS to examine how varying the value of the standard deviation $\sigma$ affects the location and shape of the graph of the probability density function.

   Fix the value of the mean $\mu$ to be 75. Graph the probability density function for several different values of $\sigma$. Summarize your observations.

2. Estimate some probabilities.

   Recall that the probability that $x$ lies between $a$ and $b$ equals

   $$P(a \le x \le b) = \int_a^b f(x)\ dx,$$

   where $f$ is the probability density function.

   Consider, once more, the probability density function where $\mu = 75$ and $\sigma = 10$.

   **a.** Estimate the probability that a score is between 70 and 80—that is, estimate the value of $P(70 \le x \le 80)$—by applying the Trapezoidal Rule with $\Delta x = 0.5$. Use ISETL to do the calculations.

   **(1)** Print a copy of your ISETL session and place it below.

   **(2)** On one of the graphs you saved from part 1, shade the region corresponding to $P(70 \le x \le 80)$. Place the graph below.

   **b.** Estimate the probability that a score is between 75 and 85—that is, estimate the value of $P(75 \le x \le 85)$—by applying the Trapezoidal Rule with $\Delta x = 0.5$. Use ISETL to do the calculations.

   **(1)** Record your estimate.

(2) Shade the region corresponding to $P(75 \leq x \leq 85)$ on the third copy of the graph which you set aside in part 1. Place the graph below.

c. Use your result from part b to estimate $P(65 \leq x \leq 75)$.

Up to this point, we have only considered continuous functions. But, what about applying the concept of integration to situations modeled by discrete functions, which are represented by scatterplots, tables, or sets of ordered pairs? If the data set is associated with a continuous function—that is, if it makes sense to "connect the dots" in the scatterplot—then the Trapezoidal Rule provides a way to approximate the associated integral. For example, in Unit 5, you used a rectangular approach to estimate the area of the Susquehanna River property. The trapezoidal approach provides a better approximation. Consider this situation again.

## Task 8-13: Using the Trapezoidal Rule on a Data Set

In Tasks 5-7 through 5-9, you examined a piece of property bounded on the north by the Susquehanna River and on the south by state Rt. 70. You wanted to determine the area of the property, which unfortunately has an irregular shape as a result of the river boundary. The seller, however, had made a careful sketch of the site boundaries as shown below.

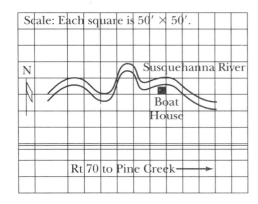

Moreover, by placing the property on a coordinate system with the southwest corner at origin and the horizontal and vertical axes determined by

Rt. 70 and the western boundary, respectively, the seller had "walked off" the distance from the highway to the river at 50-foot intervals and collected the following data:

| $x$ | 0 | 50 | 100 | 150 | 200 | 250 | 300 | 350 | 400 | 450 | 500 | 550 |
|---|---|---|---|---|---|---|---|---|---|---|---|---|
| $h(x)$ | 130 | 168 | 170 | 142 | 136 | 220 | 165 | 170 | 172 | 140 | 112 | 100 |

1. Let $h$ be the name of the function represented by the data set.

   a. Express the area of the property in terms of a definite integral.

   b. Explain why you cannot solve this integral directly.

2. Use your CAS to graph the data.

   a. Print a copy of your graph and place it in the space below.

   b. Sketch the boundary of the property on your graph and shade the region corresponding to the property.

3. Use the Trapezoidal Rule to estimate the area of the property.

   a. Cover the regions with trapezoids, where the base of each trapezoid is 50 feet. Sketch the trapezoids on your graph in part 2.

   b. Approximate the area of the property by applying the Trapezoidal Rule and using ISETL to do the calculations. Place a copy of your ISETL session below.

   Note that since the function $h$ is defined by a table, it can be represented in ISETL by a set of ordered pairs, where

$$h := \{[0, 130], [50, 168], \ldots\}.$$

In the last task, you used the Trapezoidal Rule to estimate a definite integral that was associated with a discrete data set. In the next task, you will use the same approach to tackle a project that examines the question: How big is the national debt?

First, consider some background notes for the project. In the 1992 presidential campaign in which Republican George Bush was unseated by Democrat Bill Clinton, there was a great deal of talk about reducing the federal deficit. This was stimulated by the candidacy of independent Ross Perot, whose major issue was debt reduction. Many people think that the debt and the deficit are the same thing, which is not the case. The *annual federal deficit* is the amount of money spent each year in excess of the government revenues, whereas the *national debt* is the *accumulation* of the deficits. Thus, it is possible to reduce the annual deficits and still have a national debt that increases each year. The question is this: Do the reported receipts and outlays for various years actually yield values for the national debt that are close to the reported values? If not, the federal government could be deceiving taxpayers. Because the national debt is the accumulation of annual deficits, integration techniques can be used to keep a running total of the mounting deficits. This assumes, of course, that data on government receipts and outlays is available to help determine the annual deficits.

Clinton briefs congressional leaders Tuesday. (AP)

# What's in Clinton's plan?

Clinton will propose a four-year, $500 billion deficit-reduction package roughly divided between new taxes and spending cuts. It will include a short-term, $31 billion sweetener to boost job creation, including $16 billion in highway and other public works projects and $15 billion in tax breaks for businesses to create private-sector jobs.

WASHINGTON (AP) — Clinton Admini...
year should feel some pinch under th...
But Clinton has said 70 percent ...
over $100,000.

Reproductions from the *Harrisburg PA Patriot News*, 2/13/93.

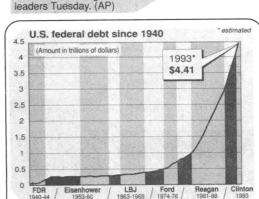

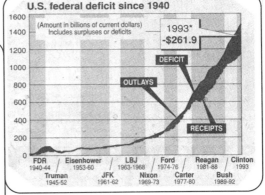

Charts compiled by Wm. J. Castello of the Associated Press from data provided by the U.S. Congressional Budget Office.

Suppose you are an investigative reporter who has been asked to verify allegations that a member of Ross Perot's political party has made with regard to the federal debt. This party member claims that during the Bush

Administration ending in 1992, the government had been using accounting tricks to make the annual deficit look smaller than it actually was. She alleges that more than $124 billion dollars of the Social Security Trust Fund and Transportation Fund money was borrowed to reduce the reported deficits in 1990 and 1991. Another alleged "trick" with the budget includes a deliberate failure to post over $100 billion in Savings and Loan bailout funds as expenditures.

If the allegations are true, then the actual federal debt should be much larger than accumulations of reported revenues and outlays would total. Your editor has given you one night to verify that there is a significant discrepancy between the national debt and the accumulation of reported deficits. The editor would like to report in the morning edition that either the allegations are plausible *or* they are unreasonable. Late that night, as you frantically try to get your hands on the data, you pull out the 46th edition of the *Information Please Almanac* published in 1993 and search through it. Yes indeed, there are tables of reported receipts and outlays as well as a table summarizing the national debt by year. On page 47, there is a Treasury Department table that lists the national debt at the end of fiscal year 1939 as $40 billion. In 1991, the debt was reported to have risen to $3665 billion. Then on pages 64 and 65, you discover a table of receipts and outlays for some of the years between 1939 and 1991. These data are summarized as follows

Federal Revenues/Outlays ($Millions)*

| Year | Revenues | Outlay |
|------|----------|--------|
| 1939 | 4,979 | 8,841 |
| 1943 | 21,947 | 79,368 |
| 1944 | 43,563 | 94,986 |
| 1945 | 44,362 | 98,303 |
| 1950 | 36,422 | 39,544 |
| 1956 | 74,547 | 70,460 |
| 1960 | 92,492 | 92,223 |
| 1965 | 116,833 | 118,430 |
| 1970 | 193,743 | 196,588 |
| 1975 | 280,997 | 326,105 |
| 1980 | 520,050 | 579,011 |
| 1985 | 733,996 | 936,809 |
| 1986 | 769,091 | 989,789 |
| 1987 | 854,143 | 1,002,147 |
| 1988 | 908,953 | 1,064,055 |
| 1989 | 990,691 | 1,142,643 |
| 1990 | 1,031,462 | 1,251,850 |
| 1991 | 1,053,832 | 1,322,561 |

If data were available for each year, you could simply calculate the deficits as the differences between outlays and revenues and add them up to find the accumulated debt. But, it is not that simple. The first thing you notice is that the data have not been reported every year. How are you go-

*From Financial Management Service (U.S. Treasury Department), *Information Please Almanac*, 46th Edition, Houghton Mifflin, Boston, 1993, pp. 64–65.

ing to be able to estimate the national debt with this incomplete data? Wait, all is not lost! You realize that you might be able to use calculus to estimate the accumulated deficits. If the deficit does not change wildly from year to year, then numerical integration might give you a reasonable estimate of the total debt. Suddenly, in desperation, you remember the Trapezoidal Rule you learned in calculus class back in college. Aha, you think to yourself, you can check out the validity of the allegations by using technology to do the calculations. Is it possible, you wonder, that the government has been deceitful? You set to work on the project.

## Task 8-14: (Project) Estimating the National Debt

1. Model the situation by representing the deficit by a function.

   a. Create a new column in the table, which lists the Federal revenues/outlays, and label it "Deficit."

   b. For each reported year from 1939 through 1991, calculate the annual deficit (in millions) by subtracting the revenues from the outlay for that year. Enter your results in the Deficit column in the table.

   c. Use your CAS to graph the annual deficit (in millions) as a function of time for the years between 1939 and 1991. Print a copy of your graph and place it in the space below.

2. Explain why integrating the deficit function between 1939 and 1991 provides an estimate of the accumulated debt during this time period.

3. Investigate the feasibility of using the trapezoidal approach to analyze this situation.

   a. Based on your graph, do the deficit amounts vary in a wildly unpredictable way during the years for which data are reported or do the deficits just seem to increase faster and faster?

   b. Consequently, do you think the Trapezoidal Rule will provide a reasonable estimate of the mounting debt or do you think there will be too much uncertainty associated with the deficits for the years between those reported?

4. Analyze the situation.

  **a.** Suppose no accounting tricks have been used by the government. Estimate the change in the debt between 1939 and 1991 by applying the Trapezoidal Rule. Do the calculations using your calculator, your CAS, or a spreadsheet package. (Note that the bases of the trapezoids differ in size. For example, the base of the first trapezoid equals 1943 minus 1939 or 4, whereas the base of the second is 1944 minus 1943 or 1.)

  Record your estimate of the accumulated deficits between 1939 and 1991, in terms of billions of dollars, below.

  **b.** Find the difference between the reported values of the national debt in 1939 of $40 billion and in 1991 of $3,665 billion.

  **c.** Compare your estimate of the accumulated deficits with the reported change in the national debt between 1939 and 1991, and answer the following questions. Be sure to justify why you feel the way you do. (Note that there are no single right answers to these questions.)

  **(1)** Do the accumulated deficits, based on the reported revenues and outlays from 1939 to 1991, account for the change in the debt during this time period? Or is it possible that revenues have been overreported and outlays underreported from year to year?

  **(2)** In view of your results and your answers to the questions in part 2, what will you tell your editor? How will you write your story?

In the last two tasks, you approximated the value of the definite integral associated with a discrete function by using the Trapezoidal Rule. If the graph of the discrete function is fairly "regular," then an alternate approach is to find a continuous function that fits the data and then integrate the curve-fit function. In the next two tasks, you will apply this approach to a real-life situation.

Suppose you campaigned tirelessly for, and contributed big bucks to, a mayoral candidate in a recent election, and your candidate won. To help repay you for all your support, she has appointed you Director of Public Works for the city. Your very first week in office you receive a terse letter from the State informing you that you must immediately cover the city reservoir, because they claim that all sorts of undrinkable "thingies" are getting in the water. When you go to write up the specifications for the job, you find that you need to know the surface area of the reservoir. No problem, you say to yourself, until you discover that no one on your staff has any idea what the surface area is and that the reservoir is oddly shaped. Fortunately, the reservoir is symmetric and you realize that if you could fit a function to half of its perimeter, you could estimate the area by integrating the fit function and doubling the result. You decide to go for it.

*Note: The notion of curve fitting was introduced in Unit 3, Section 4. Your CAS course handout also explains how to fit a curve to a data set. You should review this information before beginning Task 8-15.*

## Task 8-15: Fitting a Curve and Using the Model

The first thing you do is go out to the reservoir and collect some data. As you survey the situation, you imagine the horizontal axis dividing the reservoir in half, so that the same shape is above and below the axis. You plan to estimate the total area by finding an expression that models the upper half of the reservoir, calculating the area bounded by the function, and doubling the result. With the help of your assistant, you measure the width of the reservoir (perpendicular to your imaginary horizontal axis), at 15-foot intervals. Because you are only interested in the upper half of the reservoir, you divide each measurement in half and record the results in the table below.

| x (ft) | 0 | 15 | 30 | 45 | 60 | 75 | 90 | 105 | 120 | 135 | 150 |
|--------|---|------|------|------|------|------|------|------|------|------|------|
| y (ft) | 0 | 41.9 | 69.8 | 86.9 | 96.0 | 99.6 | 99.8 | 96.1 | 86.0 | 76.4 | 73.8 |

| x (ft) | 165 | 180 | 195 | 210 | 225 | 240 | 255 | 270 | 285 | 300 |
|--------|------|------|------|------|------|------|------|------|------|------|
| y (ft) | 76.4 | 86.0 | 96.1 | 99.8 | 99.6 | 96.0 | 86.9 | 69.8 | 41.9 | 0 |

1. Use your CAS to plot the data set.

   a. Print a copy of the graph and place it below.

   b. On the copy of the graph, sketch the perimeter of the entire reservoir, noting that your data represents only the upper half.

2. Fit a polynomial function to the data.

   a. Record the expression that best fits the data.

   b. Graph the data set and the curve-fit function on the same pair of axes. Print a copy of the graph and place it below.

3. Approximate the surface area of the upper half of the reservoir by integrating the curve-fit function over the appropriate interval. Use your CAS to evaluate the definite integral. Record the definite integral and give its value.

**4.** Approximate the total surface area of the reservoir, using your result from part 3 and the fact that the reservoir is symmetric.

**5.** Actually, to solve this problem, you did not have to do all the measuring that you initially did. Explain how you might have saved yourself (and your assistant) some time and effort.

Sometimes, instead of fitting a discrete function with a single expression, you need to fit it in pieces—that is, you need to fit different portions of the discrete function with different curves. Then, you can approximate the definite integral associated with the discrete function by integrating each curve-fit function over its particular domain and summing the results. This is the approach you will use in the next project, as you examine the question: What was the average temperature?

First, some background information. Weather conditions such as clouds, winds, rain, and snow cause the temperature patterns on each day at each location on Earth to be unique. However, on calm days when there are no clouds or storms, patterns of heating and cooling can be described with familiar mathematical functions. Residents in Carlisle, Pennsylvania enjoyed a wonderful spring week consisting of calm sunny days and nights in April 1991. A reproduction of a temperature chart along with hourly data estimated from it for part of Sunday April 9th and Monday April 10th are shown in Figure 4.

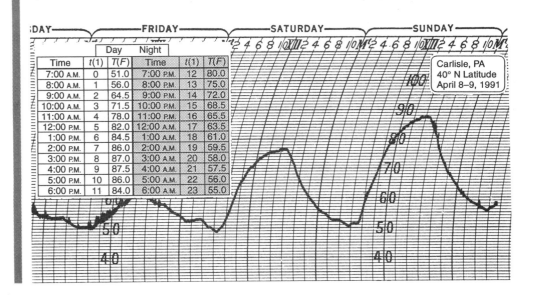

| | Day | | | Night | | |
|---|---|---|---|---|---|---|
| Time | $t(1)$ | $T(F)$ | Time | $t(1)$ | $T(F)$ | |
| 7:00 A.M. | 0 | 51.0 | 7:00 P.M. | 12 | 80.0 | |
| 8:00 A.M. | 1 | 56.0 | 8:00 P.M. | 13 | 75.0 | |
| 9:00 A.M. | 2 | 64.5 | 9:00 P.M. | 14 | 72.0 | |
| 10:00 A.M. | 3 | 71.5 | 10:00 P.M. | 15 | 68.5 | |
| 11:00 A.M. | 4 | 78.0 | 11:00 P.M. | 16 | 65.5 | |
| 12:00 P.M. | 5 | 82.0 | 12:00 A.M. | 17 | 63.5 | |
| 1:00 P.M. | 6 | 84.5 | 1:00 A.M. | 18 | 61.0 | |
| 2:00 P.M. | 7 | 86.0 | 2:00 A.M. | 19 | 59.5 | |
| 3:00 P.M. | 8 | 87.0 | 3:00 A.M. | 20 | 58.0 | |
| 4:00 P.M. | 9 | 87.5 | 4:00 A.M. | 21 | 57.5 | |
| 5:00 P.M. | 10 | 86.0 | 5:00 A.M. | 22 | 56.0 | |
| 6:00 P.M. | 11 | 84.0 | 6:00 A.M. | 23 | 55.0 | |

Carlisle, PA
40° N Latitude
April 8–9, 1991

Figure 4. A reproduction of an actual temperature chart recorded at Dickinson College during April 8–10, 1991. Temperature data shown in the table are estimated from the chart.

Consider what happens to the temperatures for the 24-hour period beginning at 7 A.M. on Sunday, April 9th. There are two distinct parts to the curve: one where the temperatures are changing as a result of daytime solar radiation and one where temperatures are changing due to nighttime radiation cooling. In the first part of Task 8-16, you will fit a function to the day portion of the graph and one to the night portion, and you will relate these functions to some theoretical understandings that scientists have about the way solar energy is gained and lost on an ideal day like April 9–10. In the second part of the task, you will estimate the average temperature for the 24-hour period first by using the crude approach which the weather service uses and then by using integration techniques which give a more accurate estimate. Finally, you will compare your results.

## Task 8-16: (Project) Finding an Average Temperature

**1.** Model the situation by fitting a curve to the temperature data.

The goal of this part of the project is to fit a function $T$ to the temperature data given in Figure 1, for the 24-hour period beginning at 7 A.M on Sunday. To get as close a fit as possible, you will do this in two pieces: first by finding a function that fits the day data for Sunday from 7 A.M. to 6 P.M.—when solar energy is being absorbed by the Earth's surface—and then by finding another function to fit the night data for Sunday from 7 P.M. to Monday at 6 A.M.—when solar energy is being lost from the Earth's surface.

**a.** To help you get oriented, compare the graph and the data table given in Figure 4.

**(1)** Look at the graph in Figure 4 carefully, observing that the hours of the day are located along the upper horizontal axis, where M represents midnight and XII noon, and the temperature in Fahrenheit—50°F, 60°F, and so on—is given on the vertical axis.

Use the graph to estimate the temperature at the following times.

**(a)** Sunday at 7 A.M.

**(b)** Sunday at noon

**(c)** Sunday at 3 P.M.

**(d)** Sunday at midnight

**(e)** Monday at 6 A.M.

**(2)** Next, look at the data table, "Day Night," in Figure 4, where $T$ is the temperature in Fahrenheit (°F) at time $t$ in hours (h). Compare the estimates which you made in part (1) to the values given on the table. Your estimates should be close to the corresponding entries in the data table. If there are any discrepancies, determine why.

**b.** Fit a function to the day data for Sunday, from 7 A.M. to 7 P.M.

Before you do this, think a little bit about what is going on here. Although the Sun is more than 90 million miles away, it provides the energy needed to sustain life on Earth and drive the weather cycles that make up the annual seasons. Solar energy consists of electromagnetic radiation of different wavelengths, including infrared light, visible light, ultraviolet light, X-rays, and gamma rays. The layers of dust, moisture, clouds, and air in the Earth's atmosphere shield us from the full effect of solar radiation. Of the radiation that reaches the outer atmosphere, about half is reflected back into space or absorbed by the atmosphere. The other half is absorbed by the surface of the Earth.

To maintain its average temperature, the Earth's surface must lose the same amount of energy that it receives in the daytime by reradiation into outer space at night, by evaporation, and by wind cooling. By drawing on what scientists already know about the processes of heating and cooling and using some common sense, it is possible to find functions that fit the daytime and nighttime portions of the temperature graph on our ideal, cloudless day and night. This is the goal of the first part of this project.

The rate at which solar energy is absorbed by the Earth's surface depends most critically on the angle between the surface of the Earth and the direction of the Sun's rays as shown in Figure 5. It can be seen from the figure that when the Sun is high overhead, its rays are more concentrated and it loses less energy by passing through the atmosphere more directly. When the Sun is low in the sky, it passes through less atmosphere and its rays are spread out more as they hit the surface of the Earth.

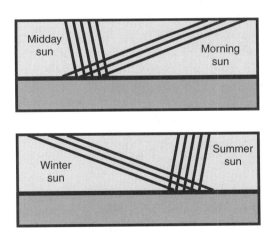

Figure 5. Diagram showing how the amount of solar energy incident on the ground increases with the angle between the Sun and the surface of the Earth because the rays from a high Sun are more concentrated and pass through less atmosphere.

This angle depends on the latitude, the time of day, and the time of year. For example in Carlisle, Pennsylvania, located at 40°N latitude, the Sun's rays subtend an angle of 73.5° relative to the surface of the Earth at noon during the summer solstice (June 21st) and an angle of only 26.5° at noon during the winter solstice (December 21st). The tilt of the Earth's axis of 23.5° relative to the plane of its orbit around the Sun is what causes this variation from season to season as shown in Figure 6.

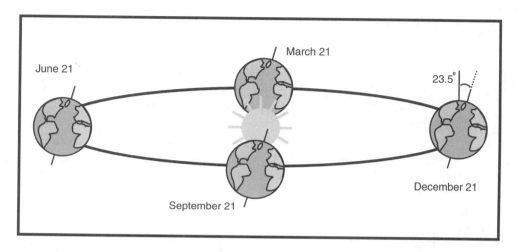

Figure 6. The Earth turns on its own axis once each day; it takes 365 days to revolve around the Sun. The relative sizes of the Earth and the Sun and the distances between them are not to scale.

The day is in the early spring when the days and nights are about the same length. When the Sun first rises, its rays are parallel to the surface of the Earth and no heating takes place. Then, since Carlisle is at 40°N latitude, the Sun rises higher in the sky until it reaches a height of 90° minus 40°, or 50°, at about noon. It then sinks to an angle of 180° at sunset. The functional relationship describing heating is fairly complicated. The change in the Sun's angle is a function of the time of day, the amount of time needed for the sunlight to cause a temperature rise, and the amount of secondary radiation coming from the atmosphere. However, it seems reasonable that the temperature starts at a low value, rises to a maximum, and then heads back for the low value after the Sun sets. In fact, the graph given in Figure 4 shows that this is exactly what happens. The arc that is made is a piece of a familiar curve.

To simplify the input, represent each hour with the corresponding value of time $t$, where $0 \leq t < 24$; that is, assume $t = 0$ corresponds to 7 A.M. on Sunday, $t = 1$ corresponds to 8 A.M. on Sunday, and so on. Fit a function to the day data, where $0 \leq t < 12$.

(1) First, try to guess the shape of the function fitting the daytime temperatures. What function approximates the changes in temperature over time of the Earth's surface in the daytime as the number of rays incident on the Earth starts at zero, rises to a maximum, and returns to zero?

(2) Use your CAS to fit the day data $(0 \le t \le 11)$ with a function called *TDay*. Enter the expression for the curve fitting the day data in the space below.

(3) According to the data table, the value of $T$ at 0 hours is 51°F. Find the difference between $T(0)$ and *TDay*$(0)$. Your answer should be very small. If it is not, try to find a more accurate model for *TDay*.

(4) Plot the day data and the function modeling the data, *TDay*, on the same pair of axes. Print a copy of the graph and place it below.

c. Fit a function to the night data from 7 P.M. Sunday to 7 A.M. Monday.

Before you try to do this, consider some more background information.

At night, energy is lost due to a process called radiation cooling. In 1879, Josef Stefan proposed that the rate at which a surface cools is proportional to the fourth power of its absolute temperature $T$. He based his proposal on experimental data. As the surface of the Earth cools, the atmosphere—which is also storing energy from the Sun—radiates some of its energy toward the Earth. Thus, according to Stefan, close to the surface of the Earth at night, the rate of change in temperature $T$ with respect to time $t$ is given by a differential equation of the form

$$\frac{dT}{dt} = c_1 T^4 - c_2 T^4_{\text{atm}},$$

where $T_{\text{atm}}$ is the average temperature of the atmosphere and $c_1$ and $c_2$ are other constants. If you know the temperature in the early evening when the cooling begins, then you can use integration to find a function that describes how the temperature $T$ changes with time $t$. In particular, since

$$dt = \frac{1}{c_1 T^4 - c_2 T^4_{\text{atm}}} \, dT,$$

all you have to do is evaluate the integral

$$\int_0^t dt = \int_0^t \frac{1}{c_1 T^4 - c_2 T^4_{atm}} \, dT.$$

Indeed this integral is pretty difficult. Based on the techniques you know so far, it can only be done by table lookup or the use of your CAS. By looking up the form of the integral listed above and making some good approximations, it is possible to theorize that the equation which approximately describes how $T$ decreases over time during the night is an exponential decay function which has the form

$$T = (T_{max} - T_{atm})e^{-\alpha(t-12)} + T_{atm},$$

where $T_{max}$ is the maximum temperature during the specified time interval.

The bottom line of this discussion is that to fit a curve to the night data, you should consider an exponential function. Try this.

(1) Recall that a function with the form $y = e^{-\alpha t}$ is an *exponential decay* function, where $\alpha$ is known as the *decay constant*. Use your CAS to plot $y = e^{-\alpha t}$ for several different values of $\alpha$ between 0.1 and 0.2, where $12 \leq t < 24$. Describe how the shape of the graphs changes as the value of the decay constant increases.

(2) Now, according to the previous discussion, the curve fitting the night data, for $12 \leq t < 24$, has the form

$$TNight = (T_{max} - T_{atm})e^{-\alpha(t-12)} + T_{atm}.$$

(a) Find the value of $T_{max}$ by looking at the data table in Figure 4.

(b) Simplify the expression for $TNight$, using the specific value of $T_{max}$ and the fact that $T_{atm} = 48.2°F$.

(3) At this point, the only unknown in the expression for $TNight$ is the value of the decay constant $\alpha$. Try to find a value for $\alpha$ so that the $TNight$ fits the night data on the data table as closely as possible. Do this by using your CAS to plot the night data and the function modeling the night data, $TNight$, on one pair of axes for several values of $\alpha$. Record your best guess for $\alpha$ below.

(4) The actual value of $\alpha$ is 0.13859. Plot the night data and the function *TNight* with $\alpha$ equal to 0.13859 on the same pair of axes. Print a copy of the graph and place it below.

d. At last, you can put together what you did in parts b and c to find a curve that models the data for the entire time period—that is, for $0 \leq t < 24$. Because you have one function modeling the data for the day and another for the night, the function *T* modeling the total time period is a piecewise-defined function. Use the expressions for *TDay* and *TNight* to represent *T* using the "curly bracket" notation.

$$T(t) = \begin{cases} \rule{4cm}{0.4pt}, & \text{if } 0 \leq t < 12 \\ \rule{4cm}{0.4pt}, & \text{if } 12 \leq t < 24. \end{cases}$$

2. Find the average daily temperature for the 24-hour period beginning at 7 A.M. on Sunday, April 9th.

Obviously, you will need to use the function *T* from part 1 which models the data for the desired time period. Before you do this, however, think a little bit about what is meant by "average" and about how you can find averages by using integration. (See HW7.20: The Mean Value Theorem for Integrals.)

Those who deal with the delivery of coal, oil, gas, and electricity to local residences and industries are interested in finding out quickly what the daily average temperature is. This helps in making decisions concerning the rate of production of energy for delivery. The weather service has a quick and easy way of calculating this average which is not always accurate. They simply average the maximum and minimum temperatures. The question is: Does the real average correspond to the average of the high and low temperature for a real day like April 9–10 in 1991 in Carlisle? Before you try to answer this, consider the average temperature on a hypothetical day that can be described by a simple function. How does the average temperature compare to that obtained using the quick and easy method? Once you answer this question for a hypothetical day, you will tackle the question for April 9th and 10th.

a. Consider a hypothetical day.

Suppose the temperature is described by the equation

$$T(t) = 29 \sin(0.20t) + 51,$$

where the average temperature was 56.5°F. A graph of this equation is shown in Figure 7.

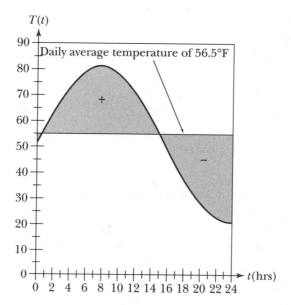

Figure 7. Graph of $T$ for the hypothetical day.

(1) Use the graph in Figure 7 to estimate the following:

 (a) The minimum temperature for the hypothetical day

 (b) The maximum temperature the hypothetical day

(2) Approximate the average temperature, using your estimates from part (1).

*Note: This is the (crude) approach that weather service uses to do the calculation.*

$$T_{\text{crude}} = \underline{\hspace{2cm}} \text{°F.}$$

(3) Look again at Figure 7, and consider the rectangle whose height is the actual temperature average $T_{\text{avg}}$, or 56.5°F, and whose width is 24 hours. When the temperature varies continuously with time, the area of this rectangle equals the area under the graph of $T$. Explain why this makes sense by considering the area over the average line (marked +) and the area under the average line (marked −).

**(4)** Equating the area of the rectangle with the area under the graph of $T$ gives

$$24 T_{avg} = \int_0^{24} T(t) \, dt.$$

Use your CAS to calculate the average temperature $T_{avg}$ for the hypothetical day, recalling that $T(t) = 29 \sin(0.20t) + 51$.

**(5)** How does the actual average temperature [from part (4)] compare to the crude average [from part (2)] for the hypothetical day?

**b.** Consider the real day, April 9th–10th.

   Suppose the temperature is properly described by the equations for day and night that you found in part 1. Now, you can check on the comparison of the crude method and the more accurate method for finding the daily average. This time, however, you are using real data.

**(1)** Determine the minimum temperature for April 9th–10th and the maximum temperature by using the data table in Figure 4.

**(2)** Use the crude weather service method to approximate the average temperature for this real day.

$$T_{crude} = \underline{\hspace{2cm}} \,°F.$$

**(3)** Express the actual average temperature for the 24-hour period starting at 7 A.M. on Sunday, April 9, as the sum of two definite integrals using the integral equation for $T_{avg}$ which you derived above in part a (4) and the piecewise representation for expression for $T$ which you derived at the end of part 1.

**(4)** Find the actual average temperature, using your CAS to evaluate the integrals. Enter the result below.

(5) How does the actual average temperature compare to the crude average for the real day?

## Unit 8 Homework After Section 4

- Complete the tasks in Section 4. Be prepared to discuss them in class.

- Write your journal entry for this unit. As usual, before you begin to write, review the material in the unit. Think about how it all fits together. Try to identify what, if anything, is still causing you trouble.

**HW8.12** Write your journal entry for Unit 8.

1.  Reflect on what you have learned in this unit. Describe in your own words the concepts that you studied and what you learned about them. How do they fit together? What concepts were easy? Hard? What were the main (important) ideas? Give some examples of the main ideas.

2.  Reflect on the learning environment for the course. Describe the aspects of this unit and the learning environment that helped you understand the concepts you studied. What activities did you like? Dislike?

# Unit 9:

# USING DIFFERENTIATION AND INTEGRATION

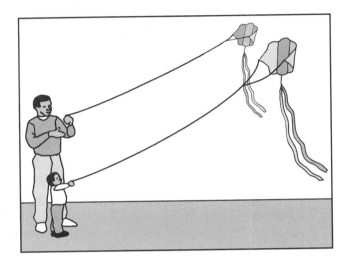

*. . . applications should motivate theory so theory is seen by students as useful and enlightening. . . .*

National Research Council, 1991
*Moving Beyond Myths: Revitalizing
Undergraduate Mathematics*

## OBJECTIVES

1. Differentiate equations implicitly.

2. Examine inverse trigonometric functions and find their derivatives.

3. Model situations using equations involving derivatives.

4. Solve related rates problems.

5. Solve separable differentiable equations.

6. Investigate an uninhibited growth model.

7. Approximate solutions to differential equations using Euler's Method.

8. Evaluate improper integrals.

9. Find volumes of solids of revolution using disks and washers.

## OVERVIEW

At this point, you know what a derivative is, you know what a definite integral is, and you how these two concepts are related. In this unit, you will apply these ideas to some other situations where differentiation and integration are useful.

In the first section, you will investigate how to use *implicit differentiation* to find the derivative of a function, where the relationship between the independent and dependent variables is represented by an equation. For example, you will discover how to find $y'$, when $x^2 + xy - y^2 = 6x$ and $y$ is a function of $x$. You will also define the *inverse trigonometric functions* and use implicit differentiation to find their derivatives.

In the next section, you will examine some equations involving derivatives. One type of situation, which occurs frequently in applications, is when you have two variables, both of which are functions of a third variable and whose relationship is represented by an equation. For instance, the area and the radius of a circle might both be functions of time where $A = \pi r^2$. Because the variables are related by an equation, their derivatives are also related by an equation. Consequently, problems of this type are called *related rates* problems. In a typical problem, you know the rate of change of one of the variables at a particular moment in time and want to determine the rate of change of the other variable at that moment.

You will consider other equations containing derivatives called *differential equations*. In this case, you have an equation that gives the relationship between an (unknown) function of a single variable, its first- and/or higher-order derivatives, and its dependent variable. For instance, $(x^2 - 1)\dfrac{dy}{dx} = xy$ and $y' = 3y$ are differential equations. The goal is to find a function that satisfies the given equation. Entire courses are devoted to this topic; the tasks in this unit will help you understand what differential equations are, why they are useful, and how to solve equations where the variables can be "separated."

In the last section, you will turn your attention from differentiation to integration. You know how to evaluate the definite integral of a continuous function, where the limits of integration are finite. For instance, you know how to evaluate $\displaystyle\int_1^2 \frac{1}{x^2}\, dx$. But what if one of the limits of integration is infinite, such as $\displaystyle\int_1^\infty \frac{1}{x^2}\, dx$? Or, what if the underlying function has a blowup discontinuity at one of the limits of integration, such as $\displaystyle\int_0^{16} \frac{1}{\sqrt{x}}\, dx$? These types of integrals are called *improper integrals*, and, surprisingly, even though the regions associated with these integrals are unbounded, an integral of this type may have a finite value. You will investigate how to determine if this is the case.

Finally, you will investigate how to find the volume of a *solid of revolution* which is generated by revolving a region in the *xy*-plane about a line, such as the *x*-axis. You will build models to help visualize what is occurring. You will find formulas for calculating the volume of a solid of revolution by applying the same approach you used to find the area of a region. Only this time, instead of covering the region with shapes whose areas you can calculate, you will slice the solid into pieces whose volumes you can calculate. In particular, you will do the following:

- Slice up the solid into pieces whose volumes you know, such as right circular cylinders or washers.
- Sum up the volumes of the little pieces to approximate the volume observing that the more pieces you use, the better the approximation.
- Take the limit of the sum to find the exact volume.
- Express the exact volume in terms of an integral.

## SECTION 1

### Implicit Differentiation and Inverse Functions

The differentiation rules assume that the independent variable is expressed as a function of the dependent variable. For example, if $y = -2x^3 + x + 1$, then the dependent variable $y$ is a function of the independent variable $x$, and you can easily show that $y' = -6x^2 + 1$. Sometimes, however, the relationship between the independent and dependent variables is represented by an equation, such as $x^2 + 2xy = y^2$, where the $x$'s and the $y$'s are intermixed. In this case, $y$ is said to be an *implicit function* of $x$. The question is how to find $y'$ without first solving the given equation for $y$.

One approach is to differentiate both sides of the equation with respect to $x$. To do this, you assume that $y$ is a function of $x$, even though you do not know what the function is. After you take the derivative of both sides, you solve the resulting equation for $y'$. This process is called *implicit differentiation*. For example, to use implicit differentiation to find the derivative of $x^2 + 2xy + y^2$, you might do the following:

$$x^2 + 2xy = y^2$$ 
Given.

$$x^2 + 2x \cdot f(x) = (f(x))^2$$ 
Assume $y = f(x)$.

$$\frac{d}{dx}(x^2 + 2x \cdot f(x)) = \frac{d}{dx}((f(x))^2)$$ 
Take the derivative of both sides with respect to $x$.

$$2x + 2xf'(x) + 2f(x) = 2f(x)f'(x)$$ 
Differentiate each term with respect to $x$.

$$f'(x) = \frac{2f(x) + 2x}{2f(x) - 2x}$$ 
Solve for $f'(x)$.

$$y' = \frac{y + x}{y - x}$$ 
By assumption $y = f(x)$. Thus, $y' = f'(x)$. Substitute for $f(x)$ and $f'(x)$.

Before starting the next task, read through this example several times. Note that when you differentiate each term with respect to $x$, you need to use the Product Rule to find the derivative of $2x \cdot f(x)$ on the left-hand side and you need to use the Extended Power Rule to find the derivative of $(f(x))^2$ on the right-hand side.

---

## Task 9-1: Using Implicit Differentiation

1. For each of the following equations, use implicit differentiation to find $y'$. Assume each equation determines an implicit function; that is, assume there exists a function $f$ such that $y = f(x)$. Do not try to solve for $y$.

   **a.** $y^4 + 3y - 4x^3 = 5x + 1$

   **b.** $3x^2 - xy + 4y^2 = 0$

   **c.** $(y^2 - 1)^2 = (2x^2 + 3x - 1)^4$

**2.** You can simplify the process of implicit differentiation by thinking of $y$ as a function if $x$—without actually substituting $f(x)$ for $y$—and using the notation $y'$ to indicate the derivative of $y$ with respect to $x$. For instance, consider once again the example preceding this task.

$$x^2 + 2xy = y^2 \qquad \text{Given.}$$

$$\frac{d}{dx}(x^2 + 2xy) = \frac{d}{dx}(y^2) \qquad \text{Take the derivative of both sides with respect to } x.$$

$$2x + 2xy' + 2y = 2yy' \qquad \text{Differentiate each term with respect to } x, \text{ keeping in mind that } y \text{ is a function of } x. \text{ Use } y' \text{ to denote the derivative of } y.$$

$$y' = \frac{y + x}{y - x} \qquad \text{Solve for } y'.$$

Consider $x^4 + 2x^2y^2 - 3xy^3 - 6x = 0$.

**a.** Use this shortcut notation to find $y'$.

**b.** Find the equation of the tangent line at $P(1,-1)$.

You can use implicit differentiation to find the derivative of the inverse trigonometric functions. First, recall what it means for a function to have an inverse. To graph the inverse of $f$, which is usually denoted by $f^{-1}$, you reflect the graph of $f$ through the line $y = x$. Frequently, however, the reflection is not a function. In particular, the reflection is not a function if $f$ is not one-to-one, in which case the graph of $f$ does not pass the Horizontal Line Test. When this happens, it may be possible to restrict $f$ to a smaller portion of its domain so that the reflection is a function over this smaller piece.

In Task 9-2, you will apply this approach to define the inverse sine, cosine and tangent functions. In Task 9-3, you will use implicit differentiation to find the derivatives of these inverse trigonometric functions.

## Task 9-2: Defining Inverse Trigonometric Functions

1. Consider the sine function.

   **a.** Sketch the graph of $f(x) = \sin(x)$, $-\infty < x < \infty$.

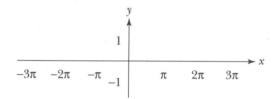

   **b.** Sketch the reflection of $f(x) = \sin(x)$ through the line $y = x$.

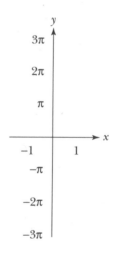

   **c.** Explain why the reflection of $f(x) = \sin(x)$ is not a function when $x$ is allowed to be any real number.

   **d.** Now, restrict the domain of the sine function to $-\dfrac{\pi}{2} \le x \le \dfrac{\pi}{2}$.

**(1)** On the axes given below, sketch the graph of the reflection of $f(x) = \sin(x)$ for $-\frac{\pi}{2} \le x \le \frac{\pi}{2}$.

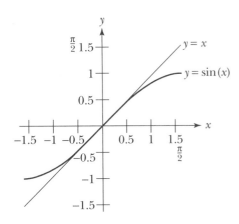

**(2)** Give the domain of the inverse sine function in this case.

**(3)** Give the range of the inverse sine function, in terms of $\pi$, in this case.

**e.** To represent the inverse of a function by an expression, you interchange the $x$'s and $y$'s and try to solve for $y$. In the case of $y = \sin(x)$, interchanging the $x$'s and $y$'s gives $\sin(y) = x$, which is impossible to solve algebraically for $y$. (Recall that this also happened when you tried to find an expression for the inverse of $y = e^x$.) Consequently, the notation $\sin^{-1}$ is used to denote the *inverse sine function*, where

$$\sin^{-1}(x) = y \quad \text{if and only if } \sin(y) = x,$$
$$\text{where } -1 \le x \le 1 \text{ and } -\frac{\pi}{2} \le y \le \frac{\pi}{2}.$$

Note that the $-1$ notation in $\sin^{-1}$ refers to the fact that this is the inverse of the sine function; it does not mean to take the reciprocal of $\sin(x)$. An alternate name for $\sin^{-1}$ is *arcsin*.

In the last part, you sketched the graph of $\sin^{-1}$. Now find some of values of $\sin^{-1}$.

**(1)** First recall some values of the sine function.

| $x$ | $-\pi/2$ | $-\pi/4$ | 0 | $\pi/4$ | $\pi/2$ |
|---|---|---|---|---|---|
| $\sin(x)$ | | | | | |

(2) Calculate some values of the inverse sine function, noting that $\sin^{-1}$ and arcsin are two names for the inverse function.

(a) $\sin^{-1}(-1)$

(b) $\arcsin(0)$

(c) $\sin^{-1}\left(\dfrac{\sqrt{2}}{2}\right)$

(d) $\arcsin\left(-\dfrac{\sqrt{2}}{2}\right)$

(e) $\arcsin(1)$

(f) $\sin\left(\sin^{-1}\left(\dfrac{\pi}{16}\right)\right)$

2. Next, consider the cosine function.

a. Explain why the reflection of the cosine function through the line $y = x$ is not a function if $x$ is allowed to be any real number. Support your explanation with a diagram.

b. Restrict the domain of the cosine function to $0 \leq x \leq \pi$.

(1) Explain why the reflection of the cosine function through the line $y = x$ is a function when the domain of $y = \cos(x)$ is restricted to $0 \leq x \leq \pi$.

(2) Sketch the graph of the inverse of $y = \cos(x)$ when $0 \leq x \leq \pi$.

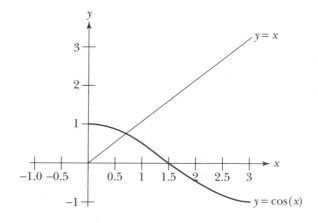

**(3)** Give the domain and the range of the inverse cosine function in this case.

**(4)** The *inverse cosine function* is defined as follows:

$$\cos^{-1}(x) = y \quad \text{if and only if } \cos(y) = x,$$
$$\text{where } -1 \le x \le 1 \text{ and } 0 \le y \le \pi.$$

An alternate name for $\cos^{-1}$ is *arccos*.

Find some values of the cosine function. Use these to determine the associated values of the inverse cosine function.

| x | cos(x) |
|---|--------|
| 0 | |
| $\frac{\pi}{2}$ | |
| $\pi$ | |

| x | $\cos^{-1}(x)$ |
|---|--------|
| 1 | |
| 0 | |
| $-1$ | |

**3.** Finally, consider the tangent function.

**a.** Recall that the tangent function is defined everywhere except at $x = \pm\pi/2$, $\pm 3\pi/2$, and so on. Explain why the reflection of the tangent function through the line $y = x$ is not a function. Support your explanation with a diagram.

**b.** Restrict the domain of the tangent function to $-\frac{\pi}{2} < x < \frac{\pi}{2}$.

**(1)** Explain why the reflection of the tangent function through the line $y = x$ is a function when the domain of $y = \tan(x)$ is restricted to $-\frac{\pi}{2} < x < \frac{\pi}{2}$.

**(2)** Sketch the graph of the inverse of $y = \tan(x)$ when $-\frac{\pi}{2} < x < \frac{\pi}{2}$.

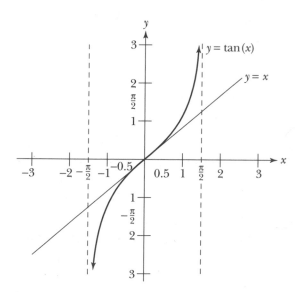

**(3)** Give the domain and the range of the inverse tangent function in this case.

**(4)** The inverse tangent function has two horizontal asymptotes. Explain why this is the case and indicate their locations on your graph in part (2).

**(5)** The *inverse tangent function* is defined as follows:

$\tan^{-1}(x) = y$   if and only if $\tan(y) = x$,
where $x$ is any real number and $-\frac{\pi}{2} < y < \frac{\pi}{2}$.

An alternate notation for $\tan^{-1}$ is *arctan*.

Use your CAS or your calculator to find some values of the inverse tangent function.

| $x$ | $-100$ | $-10$ | $-1$ | $0$ | $1$ | $10$ | $100$ |
|---|---|---|---|---|---|---|---|
| $\tan^{-1}(x)$ | | | | | | | |

You can find the derivatives of the inverse trigonometric functions by using implicit differentiation. In the next task, we will discuss how to find the derivative of the inverse sine function and then guide you through how to find the derivatives of $\cos^{-1}$ and $\tan^{-1}$.

## Task 9-3: Differentiating Inverse Trigonometric Functions

First, let us find the derivative of $\sin^{-1}(x)$. We do not know how to do this directly. However, by definition,

$$\sin^{-1}(x) = y \quad \text{if and only if } \sin(y) = x.$$

Thus, you can differentiate $\sin(y) = x$ implicitly as follows:

$$y = \sin^{-1}(x) \qquad \text{Given.}$$

$$\sin(y) = x \qquad \text{Convert.}$$

$$\frac{d}{dx}(\sin(y)) = \frac{d}{dx}(x) \qquad \text{Take the derivative of both sides with respect to } x.$$

$$\cos(y) \cdot y' = 1 \qquad \text{Differentiate both sides, keeping in mind that } y \text{ is a function of } x.$$

$$y' = \frac{1}{\cos(y)} \qquad \text{Solve for } y'.$$

$$\frac{d}{dx}(\sin^{-1}(x)) = \frac{1}{\cos(y)} \qquad \text{Because } y = \sin^{-1}(x),\ y' = \frac{d}{dx}(\sin^{-1}(x)).$$
$$\text{Substitute for } y'.$$

One disadvantage of implicit differentiation is that the derivative may contain an expression for $y$, which is true in this case. As a result, to represent the derivative of $\sin^{-1}(x)$ in terms of $x$, we need to represent $\cos(y)$ in terms of $x$. We can do this using a familiar trig identity and the assumption that $\sin(y) = x$.

$$\cos^2(y) + \sin^2(y) = 1 \qquad \text{Trig identity.}$$

$$\cos(y) = \sqrt{1 - \sin^2(y)} \qquad \text{Solve for } \cos(y).$$

$$\cos(y) = \sqrt{1 - x^2} \qquad \text{Now } \sin(y) = x, \text{ since}$$
$$\sin^{-1}(x) = y. \text{ Substitute}$$
$$x \text{ for } \sin(y).$$

Combining these two results gives

$$\frac{d}{dx}(\sin^{-1}(x)) = \frac{1}{\cos(y)} = \frac{1}{\sqrt{1 - x^2}} \quad \text{for } -1 < x < 1.$$

Now, you try it. Read once more through the derivation of the formula for finding the derivative of the inverse sine function, and then find the derivatives of the inverse cosine and tangent functions by mimicking the approach.

1. Find the derivative of the inverse cosine function, where

$$\cos^{-1}(x) = y \quad \text{if and only if } \cos(y) = x.$$

    **a.** First, show that

$$\frac{d}{dx}(\cos^{-1}(x)) = \frac{1}{\sin(y)}$$

by differentiating $\cos(y) = x$ implicitly as you fill in the missing steps in the following proof:

$$y = \cos^{-1}(x) \qquad \text{Given.}$$

$$\cos(y) = x \qquad \text{Convert.}$$

Take the derivative of both sides with respect to $x$.

Differentiate both sides, keeping in mind that $y$ is a function of $x$.

Solve for $y'$.

$$\frac{d}{dx}(\cos^{-1}(x)) = -\frac{1}{\sin(y)} \qquad \text{Because } y = \cos^{-1}(x),\ y' = \frac{d}{dx}(\cos^{-1}(x)).$$
$$\text{Substitute for } y'.$$

**b.** Next, show that

$$\sin(y) = \sqrt{1 - x^2}$$

by using a familiar trig identity and the assumption that $\cos(y) = x$.

**c.** Finally, show that

$$\frac{d}{dx}(\cos^{-1}(x)) = -\frac{1}{\sqrt{1-x^2}} \quad \text{for } -1 < x < 1$$

by combining the results from parts a and b.

**d.** Observe that neither the inverse sine function nor the inverse cosine function is differentiable at the endpoints of its domain, namely at $x = -1$ and $x = 1$. Explain why $\sin^{-1}$ and $\cos^{-1}$ are not differentiable at these values by considering their graphs. What happens to the slope of the tangent line as $x$ approaches $-1$ (from the right)? What happens as $x$ approaches 1 (from the left)?

**2.** Find the derivative of the inverse tangent function, where

$$\tan^{-1}(x) = y \quad \text{if and only if } \tan(y) = x.$$

**a.** First, show that

$$\frac{d}{dx}(\tan^{-1}(x)) = \frac{1}{\sec^2(y)}$$

by differentiating $\tan(y) = x$ implicitly.

**b.** Next, show that

$$\sec^2(y) = 1 + x^2$$

by using the trig identity $\tan^2(y) + 1 = \sec^2(y)$ and the assumption that $\tan(y) = x$.

**c.** Finally, show that

$$\frac{d}{dx}(\tan^{-1}(x)) = \frac{1}{1+x^2}$$

by combining the results from parts a and b.

**3.** To summarize:

$$\frac{d}{dx}(\sin^{-1}(x)) = \frac{1}{\sqrt{1-x^2}} \quad \text{provided } -1 < x < 1.$$

$$\frac{d}{dx}(\cos^{-1}(x)) = -\frac{1}{\sqrt{1-x^2}} \quad \text{provided } -1 < x < 1.$$

$$\frac{d}{dx}(\tan^{-1}(x)) = \frac{1}{1+x^2}.$$

Thus, by the Chain Rule,

$$\frac{d}{dx}(\sin^{-1}(f(x))) = \frac{1}{\sqrt{1-(f(x))^2}}f'(x) \quad \text{provided } -1 < f(x) < 1.$$

$$\frac{d}{dx}(\cos^{-1}(f(x))) = -\frac{1}{\sqrt{1-(f(x))^2}}f'(x) \quad \text{provided } -1 < f(x) < 1.$$

$$\frac{d}{dx}(\tan^{-1}(f(x))) = \frac{1}{1+(f(x))^2}f'(x).$$

Use these rules to find the derivatives of the following functions.

**a.** $y = \sin^{-1}(3x)$

**b.** $s = \tan^{-1}\left(\frac{1}{2}t^2\right)$

**c.**  $h(r) = r^2 \cos^{-1}(r)$

**d.**  $g(\theta) = \dfrac{\sin^{-1}\left(\dfrac{\theta}{3}\right)}{\theta + 3}$

In the last task, you used implicit differentiation to find the derivatives of three inverse trigonometric functions. In the next task, you will develop a general formula for finding the derivative of the inverse of a function.

You will also consider how to determine if a function has an inverse— that is, if its reflection through the line $y = x$ is a function. You know that $f$ has an inverse if $f$ is one-to-one, in which case the graph of $f$ passes the Horizontal Line Test. You also know that a continuous function has an inverse if it is strictly increasing or strictly decreasing, except possibly at some isolated points where the graph may momentarily level off. Consequently, a differentiable function has an inverse if its derivative is always negative or always positive, except possibly at some isolated points where $f'(x) = 0$. You will explore these ideas in Task 9-4.

## Task 9-4: Finding the Derivative of an Inverse Function

**1.** Suppose a differentiable function $f$ has an inverse function $f^{-1}$. Show that

$$\frac{d}{dx}(f^{-1}(x)) = \frac{1}{f'(f^{-1}(x))}, \quad \text{provided } f'(f^{-1}(x)) \neq 0.$$

**a.**  Since $f$ and $f^{-1}$ are inverse functions, $f \circ f^{-1}(x) = x$. Differentiate both sides of the equation with respect to $x$.

*Note: Use the Chain Rule to find the derivative of the left side of the equation, recalling that* $\dfrac{d}{dx}(f \circ g(x)) = f'(g(x))g'(x)$.

**b.** Solve your equation in part a for $\frac{d}{dx}(f^{-1}(x))$.

*Note: You should get $\frac{d}{dx}(f^{-1}(x)) = \frac{1}{f'(f^{-1}(x))}$, which is what you wanted to show.*

**2.** Consider a function that has an inverse. Show that the new formula works by finding the derivative of the inverse function two ways. First, use the direct approach—that is, represent the inverse by an expression and find its derivative. Next, use the formula from part 1. Then, show that the two approaches lead to equivalent results.

Consider $f(x) = x^2 - 3$, where $x \geq 0$.

**a.** Show that $f^{-1}(x) = \sqrt{x + 3}$, when $x \geq -3$.

**b.** Sketch graphs of $f$ and $f^{-1}$ on the same pair of axes.

**c.** Find $\frac{d}{dx}(f^{-1}(x))$ directly—that is, find $\frac{d}{dx}(\sqrt{x + 3})$.

**d.** Find $\frac{d}{dx}(f^{-1}(x))$ by applying the formula for the derivative of the inverse of a function, namely,

$$\frac{d}{dx}(f^{-1}(x)) = \frac{1}{f'(f^{-1}(x))}$$

as follows:

**(1)** Find $f'(x)$, where $f(x) = x^2 - 3$.

**(2)** Find $\dfrac{1}{f'(f^{-1}(x))}$, where $f^{-1}(x) = \sqrt{x + 3}$.

**e.** Show that your results from parts *c* and *d* are equivalent—that is, show that finding $\dfrac{d}{dx}(f^{-1}(x))$ directly and finding it using the formula lead to the same result.

**f.** Explain why the domain of $\dfrac{d}{dx}(f^{-1}(x))$ is the set of all real numbers greater than $-3$.

**3.** Suppose $f$ is a differentiable function. The reflection of $f$ through the line $y = x$ is a function if $f$ is one-to-one, in which case, the graph of $f$ passes the Horizontal Line Test. Another way to assure that the reflection is a function is to show that $f'(x)$ is always positive or that $f'(x)$ is always negative, except possibly at some isolated points where $f'(x) = 0$.

**a.** Use a graphical argument to explain why it is reasonable to assume that the reflection of $f$ is a function if $f'(x)$ is always positive, except possibly at some isolated points where $f'(x) = 0$.

**b.** Similarly, use a graphical argument to explain why it is reasonable to assume that the reflection of $f$ is a function if $f'(x)$ is always negative, except possibly at some isolated points where $f'(x) = 0$.

c. Consequently, you can determine if a differentiable function has an inverse—that is, given $f$, you can determine if $f^{-1}$ exists—by considering the sign of $f'(x)$. If either $f'(x) > 0$ or $f'(x) < 0$, except possibly at some isolated points where $f'(x) = 0$, then you know $f^{-1}$ exists, even though you might not know what $f^{-1}$ is.

Determine which of the following functions has an inverse by considering the sign of its first derivative.

(1) $f(x) = x^5 + 4x^3 - 10$

(2) $h(r) = -2.5r + 6.9$

(3) $g(t) = t^3 - 12t$

4. Consider $f(x) = x^5 + 4x^3 - 10$, which you examined above. You should have concluded that $f^{-1}$ exists since $f$ is strictly increasing. Moreover, observe that the point $P(-5,1)$ is on the graph of $f^{-1}$, since $f(1) = -5$.

Use the formula for finding the derivative of the inverse of a function, $\dfrac{d}{dx}(f^{-1}(x)) = \dfrac{1}{f'(f^{-1}(x))}$, to find the slope of the tangent line to the graph of $f^{-1}$ at $P(-5,1)$.

Note: You do not need to know the expression for $f^{-1}(x)$. All you need to know is that $f^{-1}$ exists and $f^{-1}(-5) = 1$. In other words, the formula enables you to find the value of the derivative of the inverse at a point without knowing the general expression for $f^{-1}(x)$.

5. The formula $\dfrac{d}{dx}(f^{-1}(x)) = \dfrac{1}{f'(f^{-1}(x))}$ holds if $f'(f^{-1}(x)) \neq 0$. Consequently, if $f'(f^{-1}(b)) = 0$ for $x = b$, then $\dfrac{d}{dx}(f^{-1}(b))$ does not exist. Examine what happens graphically in this case.

Suppose $f$ is strictly increasing, where $f'(x) > 0$ for all $x$, except at $x = c$, where $f'(c) = 0$. Suppose $f(c) = b$, where the graph of $f$ is given below.

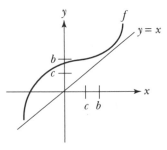

**a.** Explain why $f^{-1}$ exists.

**b.** Find $f^{-1}(b)$.

**c.** On the graph, carefully sketch:
  (1) the graph of $f^{-1}$.
  (2) the tangent line to the graph of $f$ at $x = c$.
  (3) the tangent line to the graph of $f^{-1}$ at $x = b$.

**d.** Based on the graph, explain why $f^{-1}$ is not differentiable at $x = b$.

**e.** Explain why you cannot use the formula to calculate $\dfrac{d}{dx}(f^{-1}(b))$.

## Unit 9 Homework After Section 1

- Complete the tasks in Section 1. Be prepared to discuss them in class.

- Consider some implicit functions in HW9.1.

**HW9.1** Implicit functions.

1. Given an equation, $y$ is said to be an *implicit function* of $x$, or $f$ is said to *be implicitly defined* by the equation, if substitution of $f(x)$ for $y$ leads to an identity. For example, given the equation $x^2 + 2xy = y^2$, you can use the quadratic formula to solve for $y$ and show that $y = (1 + \sqrt{2})x$ or $y = (1 - \sqrt{2})x$. Both of these functions are implicit functions of $x$, since

substituting either one into the original equation leads to an identity. For instance:

$$x^2 + 2xy = y^2 \qquad \text{Given equation.}$$

$$x^2 + 2x((1 + \sqrt{2})x) = ((1 + \sqrt{2})x)^2 \qquad \text{Substitute } (1 + \sqrt{2})x \text{ for } y.$$

$$3x^2 + 2\sqrt{2}x^2 = 3x^2 + 2\sqrt{2}x^2 \qquad \text{Expand and simplify both sides.}$$

**a.** Consider the equation $x^2 + y^2 = 1$. Find two implicit functions determined by the equation and state the domain of each function. Graph the two implicit functions.

**b.** Find at least one implicit function for each of the following equations. State the domain of the function.

(1) $\sqrt{x} + \sqrt{y} = 1$          (2) $2x^2 + 11xy - 6y^2 = 0$

(3) $x^3 - xy = 1 - 4y$          (4) $y^2 + 2x^2 = 8$

**c.** Show that the equation $x^2 + y^2 + 9 = 0$ does not determine a function $f$ such that $y = f(x)$.

**2.** Assume that each of the following equations determines a function $f$ such that $y = f(x)$. Find $y'$.

**a.** $2x^2 + 3y^2 = 10$          **b.** $x = 3y^3 - 2y^2 + y$

**c.** $xy = 9$          **d.** $\sqrt{x} + \sqrt{y} = 4$

**e.** $4 - 5xy = (y^2 + 1)^5$          **f.** $(y^2 - 4)^2 = (4x^2 + 4x + 1)^2$

**3.** Find the equation of the tangent line to the graph of $x^3 + xy + y^4 = 19$ at the point $P(1,2)$.

• Examine some inverse trigonometric functions in HW9.2.

## HW9.2

**1.** Find the value of each of the following expressions.

**a.** $\tan^{-1}(\cos(0))$          **b.** $\cos(\sin^{-1}(0))$

**c.** $\sin\left(\cos^{-1}\left(\frac{\sqrt{2}}{2}\right)\right)$          **d.** $\cos^{-1}(\cos(0.123))$

**e.** $\sin(2\arctan(1))$          **f.** $\arccos(\sin(0) - \arctan(-1))$

**2.** Use the rules for differentiating inverse trigonometric functions to find the derivatives of the following functions.

**a.** $s(t) = \cos^{-1}(3t - 6)$          **b.** $y = \arcsin\left(\frac{x}{3}\right)$

**c.** $y = x^2 \tan^{-1}(2x)$          **d.** $h(x) = \dfrac{10\tan^{-1}(x)}{x}$

**e.** $f(x) = e^{\arcsin(x^3)}$          **f.** $q(t) = \ln(\arccos(5t))$

**g.** $y = (\arctan(3x))^{10}$          **h.** $f(r) = \sqrt{\arccos(\sqrt{r})}$

**3.** Assume that each of the following equations determines a function $f$ such that $y = f(x)$. Find $y'$.

**a.** $x^3 + x\sin^{-1}(y) = xy$          **b.** $\ln(x + y) = \tan^{-1}(x)$

4. Find the slope of the tangent line to the graph of $y = \cos^{-1}(x - 1)$ at the point $P(1, \frac{\pi}{2})$.

5. Find the $x$-coordinate of each point on the graph of $y = \tan^{-1}(2x)$ where the tangent line to the graph is parallel to the line $3y - 2x + 5 = 0$.

• Examine the integration rules that correspond to the rules for differentiating inverse trigonometric functions in HW9.3.

**HW9.3** As you know, each differentiation rule has a corresponding integration rule. In particular, the rules for differentiating the inverse trigonometric functions, which you considered in this section, give you the following rules:

$$\int \frac{1}{\sqrt{1 - u^2}}\, du = \sin^{-1}(u) + C \quad \text{or} \quad -\cos^{-1}(u) + C,$$

$$\int \frac{1}{1 + u^2}\, du = \tan^{-1}(u) + C.$$

1. Evaluate the following integrals, using the method of substitution, in conjunction with the new rules.

a. $\int \dfrac{1}{1 + 25x^2}\, dx$      *Hint:* Let $u = 5x$.

b. $\int \dfrac{e^x}{\sqrt{1 - e^{2x}}}\, dx$      *Hint:* Let $u = e^x$.

c. $\int \dfrac{\sin(x)}{1 + \cos^2(x)}\, dx$      *Hint:* Let $u = \cos(x)$.

d. $\int \dfrac{1}{x^2 + 9}\, dx$      *Hints:* Multiply the numerator and denominator by $\frac{1}{9}$. Let $u = \frac{1}{3}x$.

e. $\int_0^{\sqrt{2}/2} \dfrac{x}{\sqrt{1 - x^4}}\, dx$      f. $\int \dfrac{x^2}{x^6 + 1}\, dx$

g. $\int_0^{\pi/4} \dfrac{\cos(x)}{\sqrt{1 - \sin^2(x)}}\, dx$      h. $\int \dfrac{5}{25 + x^2}\, dx$

2. Find the area under the graph of $h(t) = \dfrac{4}{\sqrt{16 - t^2}}$ for $-2\sqrt{2} \le t \le 2\sqrt{2}$.

• Consider some inverse functions in HW9.4.

**HW9.4** Recall that $\dfrac{d}{dx}(f^{-1}(x)) = \dfrac{1}{f'(f^{-1}(x))}$ is the formula for the derivative of the inverse of a function.

1. For each of the following functions:

     i. Find $f^{-1}$ and state its domain.

     ii. Find $\dfrac{d}{dx}(f^{-1}(x))$ directly.

    **iii.** Find $\dfrac{d}{dx}(f^{-1}(x))$ using the formula for the derivative of the inverse of a function.

    **iv.** State the domain of $\dfrac{d}{dx}(f^{-1}(x))$.

    **a.** $f(x) = 4 - x^2$, where $x \geq 0$     **b.** $f(x) = \dfrac{1}{x}$, where $x > 0$

    **c.** $f(x) = \sqrt{2x + 3}$, where $x \geq -\dfrac{3}{2}$   **d.** $f(x) = \sqrt{-x}$, where $x \leq 0$

2. For each of the following functions:

    **i.** Show that the function has an inverse by showing that the given function is strictly increasing or strictly decreasing. Do not try to find the inverse.

    **ii.** Use the formula for the derivative of the inverse of a function to find the slope of the tangent line to the graph of $f^{-1}$ at the indicated point $P(a,b)$.

    **a.** Consider $f(x) = -x^3 + 2$. Show $f$ has an inverse. Find the slope of the tangent line to the graph of $f^{-1}$ at $P(-25,3)$.

    **b.** Consider $f(x) = x^5 + 2x^3 + x - 10$. Show $f$ has an inverse. Find the slope of the tangent line to the graph of $f^{-1}$ at $P(-14,-1)$.

    **c.** Consider $f(x) = e^{2x} + 6$. Show $f$ has an inverse. Find the slope of the tangent line to the graph of $f^{-1}$ at $P(7,0)$.

3. Recall that the exponential function $e^x$ is the inverse of the natural logarithmic function $\ln(x)$. Use the formula for finding the derivative of an inverse function to show that $\dfrac{d}{dx}(e^x) = e^x$. *Hints:* Let $f(x) = \ln(x)$, where $f^{-1}(x) = e^x$. Assume that you know the formula for finding the derivative of the natural logarithmic function.

• Investigate the use of logarithmic differentiation in HW9.5.

**HW9.5** Given $y = f(x)$, where $f$ involves complicated quotients, products, or powers, it is sometimes useful to use a process called *logarithmic differentiation* to find the derivative of $f$. The general idea is to:

 **i.** Take the natural logarithm of both sides.

 **ii.** Simplify the results using the rules for logarithms, namely,

    $\log ab = \log a + \log b$

    $\log \dfrac{a}{b} = \log a - \log b$

    $\log a^n = n \log a$

**iii.** Differentiate implicitly, recalling that $\dfrac{d}{dx}(\ln y) = \dfrac{y'}{y}$.

**iv.** Solve for $y'$.

 **v.** Substitute $f(x)$ for $y$ on the right-hand side.

For example, to use logarithmic differentiation to find the derivative of $y = \dfrac{(5x + 9)^3}{2x^2 - 1}$, you would do the following:

$$y = \frac{(5x + 9)^3}{2x^2 - 1}$$ 
Given.

$$\ln(y) = \ln\left(\frac{(5x + 9^3)}{2x^2 - 1}\right)$$ 
Take the natural log of both sides.

$$\ln(y) = 3 \ln(5x + 9) - \ln(2x^2 - 1)$$ 
Simplify.

$$\frac{d}{dx}(\ln(y)) = \frac{d}{dx}(3 \ln(5x + 9) - \ln(2x^2 - 1))$$ 
Take the derivative of both sides.

$$\frac{y'}{y} = 3\frac{5}{5x + 9} - \frac{4x}{2x^2 - 1}$$ 
Differentiate implicitly.

$$y' = y\left(\frac{15}{5x + 9} - \frac{4x}{2x^2 - 1}\right)$$ 
Solve for $y'$.

$$y' = \left(\frac{(5x + 9)^3}{2x^2 - 1}\right)\left(\frac{15}{5x + 9} - \frac{4x}{2x^2 - 1}\right)$$ 
Substitute $\dfrac{(5x + 9)^3}{2x^2 - 1}$ for $y$.

1. Use logarithmic differentiation to find the derivatives of the following functions:

   **a.** $y = (5x + 3)^4(9x + 1)^2$  
   **b.** $y = \dfrac{(2x + 3)^{2/3}}{\sqrt{x + 1}}$

   **c.** $y = \sqrt{6x - 9}(x - 3)^5$  
   **d.** $y = \dfrac{\sqrt[5]{x^3 - 2x^2 + 6}}{\sqrt{x + 1}}$

   **e.** $y = (x - 1)^2(x - 2)^3(x - 3)^4$  
   **f.** $y = \sqrt{(2x^2 + 3)\sqrt{4x - 5}}$

2. Use logarithmic differentiation to show that:

   **a.** $\dfrac{d}{dx}(x^x) = x^x(1 + \ln(x))$  
   **b.** $\dfrac{d}{dx}(e^x) = e^x$

---

## SECTION 2

## Equations Involving Derivatives

In applications, it is not unusual for two variables, say $x$ and $y$, to be related by an equation, such as $x^2 + xy = y^2$. If one variable is a function of the other variable, say $y$ is a function of $x$, then you can use implicit differentiation to find the derivative, $y'$. Frequently, however, the variables in an equation modeling a situation are both functions of a third variable, say $t$. In this case, differentiating the given equation with respect to $t$ using the Chain Rule gives an equation that involves the rate of change of $x$ with respect to *t and* the rate of change of *y* with respect to *t*. For instance, in this case:

$$x^2 + xy = y^2 \qquad \text{Given.}$$

$$\frac{d}{dt}(x^2 + xy) = \frac{d}{dt}(y^2) \qquad \text{Take the derivative of both sides with respect to } t.$$

$$2x\frac{dx}{dt} + x\frac{dy}{dt} + y\frac{dx}{dt} = 2y\frac{dy}{dt} \qquad \text{Differentiate implicitly, keeping in mind that } x \text{ and } y \text{ are both functions of } t.$$

Observe that the relationship between the variables *x* and *y* in the original equation leads to the relationship between their rates of change in the second equation. Consequently, $\dfrac{dx}{dt}$ and $\dfrac{dy}{dt}$ are called *related rates*.

In a typical related rates problem, you are given information about the rate of change of one variable at a particular moment and asked to find the rate of change of the other variable at that moment. For example, suppose on a still evening, you toss a stone into a pond, creating a circular ripple. If the rate of change in the radius of the disturbed region is 3 feet per second, what is the rate of change of the area when the radius equals 5 feet? In this case, the area of the region, *A*, and its radius, *r*, are both functions of time, *t*, where their relationship is represented by the equation $A = \pi r^2$.

You are given that $\dfrac{dr}{dt} = 3$ ft/sec, and the objective is to find $\dfrac{dA}{dt}$ at the instant when *r* = 5 feet. Since $A = \pi r^2$, the rates of change of the area and the radius are related by the equation

$$\frac{dA}{dt} = 2\pi r\frac{dr}{dt}.$$

Substituting $\dfrac{dr}{dt} = 3$ and *r* = 5, you obtain

$$\frac{dA}{dt} = 2\pi(5)(3) = 30\pi \text{ ft/sec,}$$

which tells you that the area of the disturbed region is changing at approximately 94.2 ft/sec, when the radius of the region is 5 feet.

In general, to solve a related rates problem, you should:

- Represent the situation by a diagram. Carefully label the diagram.
- List the facts that are given. State the objective of the problem.
- Represent the relationship between the variables by an equation.
- Differentiate the equation to find a general formula that relates the rates of change.
- Solve for the general formula for the unknown rate of change.
- Consider the special case: substitute the known values.

In the next task, we will guide you through solving two typical related rates problems. In HW9.6, you will tackle some on your own.

## Task 9-5: Solving Related Rates Problems

**1.** A ladder 20 feet long leans against a vertical wall. If the bottom of the ladder slides away from the wall at the rate of 2 ft/sec, how fast is the ladder sliding down the wall when the top of the ladder is 12 feet above the ground?

> **a.** Sketch a diagram representing this situation. Label it carefully.
>
>
>
> Let $x$ denote the distance from the top of the ladder to the base of the wall (at ground level) at time $t$. Let $y$ denote the distance from the base of the wall to the bottom of the ladder at time $t$. Note that as time passes, the values of $x$ and $y$ change, whereas the length of the ladder remains fixed at 20 feet. Depict the situation on the following diagram.

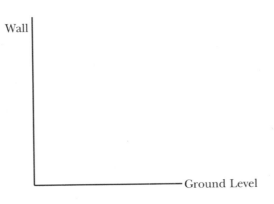

> **b.** List the facts that are given. State the objective of this problem.
>
> > **(1)** The bottom of the ladder slides away from the wall at the rate of 2 ft/sec—that is, the rate of change of $y$ with respect to $t$ is 2. Express this fact symbolically in terms of the variables.
> >
> > **(2)** The goal of this problem is to determine how fast the ladder is sliding down the wall when the top of the ladder is 12 feet above the ground. Express this symbolically.

> **c.** Represent the relationship between the variables by an equation.
>
> The variables in this case are $x$ and $y$. Determine the relationship between $x$ and $y$ by applying the Pythagorean Theorem and noting that the length of the ladder is fixed at 20 feet.

> **d.** Represent the relationship between the rates of change of the variables by an equation.
>
> Differentiate the equation in part c with respect to $t$.

**e.** Solve the general formula from part d for the unknown rate of change, which in this case is $\dfrac{dx}{dt}$.

**f.** Consider the specific case.

The goal is to find $\dfrac{dx}{dt}$ when $x = 12$ feet. In the last part, you should have discovered that

$$\frac{dx}{dt} = -\frac{y}{x}\frac{dy}{dt}$$

if you did not, go back and check your work before you continue. You were given that $\dfrac{dy}{dt} = 2$ ft/sec. You are interested in the specific case when $x = 12$ feet. To be able to calculate $\dfrac{dx}{dt}$, you need to know the value of $y$ corresponding to $x = 12$.

**(1)** Use the equation that gives the relationship between $x$ and $y$ (see part c) to find the value of $y$ when $x = 12$.

**(2)** Calculate $\dfrac{dx}{dt}$ at the specified moment.

**2.** At 11:30 A.M., the *Whisper* is 25 miles due south of the *Courageous*, on the Chesapeake Bay. If the *Whisper* is sailing west at a rate of 16 miles per hour (mi/hr) and the *Courageous* is sailing south at the rate of 20 mi/hr, find the rate at which the distance between the two ships is changing at noon.

**a.** Sketch a diagram representing the situation. Label it carefully.

Let $x$ denote the distance covered by the *Whisper* and let $y$ denote the distance covered by the *Courageous* in $t$ hours after 11:30 A.M. Let $z$ denote the distance between the two ships at $t$ hours after 11:30 A.M. In the diagram below, let $W$ and $C$ indicate the starting positions of the *Whisper* and the *Courageous*, respectively. Label the diagram.

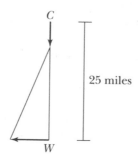

C

25 miles

W

**b.** List the facts that are given. State the objective of this problem.

   **(1)** The *Whisper* is sailing west at a rate of 16 mi/hr and the *Courageous* is sailing south at the rate of 20 mi/hr. Express these facts symbolically.

   **(2)** The goal is to determine the rate at which the distance between the two ships is changing at noon. Express this symbolically.

**c.** Note that in this problem, there are three variables, $x$, $y$, and $z$, each of which is a function of $t$. Represent the relationship among the variables by an equation.

   The variables in this case are $x$, $y$, and $z$. Determine the relationship among the variables by applying the Pythagorean Theorem.

**d.** Represent the relationship between the rates of change of the variables by an equation.

   Differentiate the equation in part c with respect to $t$.

**e.** Solve the general formula for the unknown rate of change, which in this case is $\dfrac{dz}{dt}$.

   Solve the equation from part d for $\dfrac{dz}{dt}$.

**f.** Consider the specific case.

   The goal is to find $\dfrac{dz}{dt}$ when $t = 1/2$ hour. In the last part, you should have discovered that

$$\frac{dz}{dt} = \frac{x}{z}\frac{dx}{dt} - \frac{25 - y}{z}\frac{dy}{dt}$$

   if you did not, go back and check your work before you continue. You were given that $\dfrac{dx}{dt} = 16$ mi/hr and $\dfrac{dy}{dt} = 20$ mi/hr. To be able to calculate $\dfrac{dz}{dt}$, you need to know the values of $x$, $y$, and $z$ when to $t = 1/2$ hour.

**(1)** Use the facts that $\dfrac{dx}{dt} = 16$ mi/hr and $\dfrac{dy}{dt} = 20$ mi/hr to calculate the value of $x$ and the value of $y$ when $t = 1/2$ hour.

**(2)** Use the equation that gives the relationship among $x$, $y$, and $z$ (see part c) to find the value of $z$ when to $t = 1/2$ hour.

**(3)** Calculate $\dfrac{dz}{dt}$ at noon.

In a related rates problem, you have two or more functions of a single variable, where the functions are related by an equation and, as a result, their derivatives are related by an equation. Another type of equation that involves derivatives is a differential equation. An *ordinary differential equation* or *ODE* is an equation that involves the first derivative and/or higher-order derivatives of an unknown function of a single variable. The goal is to find the unknown function. Differential equations can be used to model many real-life situations, such as the growth of human populations, the growth of yeast cells, and the rate your bank balance grows when interest is compounded continuously.

You have seen ODEs before. For instance, $y' = 2x$ is an ODE, which you can solve by finding the antiderivative of $2x$. ODEs, however, can be much more complicated. For example, if $y$ is a function of $x$, then the following are ODEs.

$$3\,\frac{d^2y}{dx^2} - x^2\,\frac{dy}{dx} - xy = 0,$$

$$\left(\frac{d^4y}{dx^4}\right)^2 - 4 = x^2\,\frac{dy}{dx}.$$

The *order* or the *degree* of an ODE is the highest-order derivative in the equation. For instance, $y' = 2x$ is a first-order ODE, whereas $\left(\dfrac{d^4y}{dx^4}\right)^2 - 4 = x^2\,\dfrac{dy}{dx}$ is a fourth-order ODE. The goal is to find the unknown function $y$. Entire courses are devoted to how to do this! One good thing about the process is that you can check your solution by substituting it into the given ODE and showing that you get an identity—that is, by showing that both sides of the equation are equal.

In the next task, you will examine what a differential equation is.*

---

*For a hands-on approach to the study of differential equations, see *A Guided Tour of Differential Equations using Computer Technology*, by Skidmore and Hale, Prentice-Hall, Englewood Cliffs, NJ, 1998. The tasks in this section are based on exercises in this book.

---

## Task 9-6: Investigating Differential Equations

When trying to solve an ODE, you start with the ODE and try to find functions that satisfy the given equation. Before trying to solve some differential equations, examine ways of representing ODEs symbolically and verbally, and think about how to convert back and forth between the representations.

**1.** Find the ODE associated with each of the following functions. Assume $y$ is a function of $x$.

| Function | Differential Equation |
|---|---|
| $y = e^{-2x}$ | $\dfrac{dy}{dx} = -2e^{-2x}$ |
| $y = mx + b$, where $m$ and $b$ are constants | |
| $x^2 + y^2 = 1$ | |
| $y = 0.05e^x$ | |
| $10 = xy$ | |
| $y = e^x + 1$ | |
| $y = \frac{1}{3}x^3$ | |
| $y = \sqrt{2x - 6}$ | |
| $y = kx^2$, where $k$ is a constant. | |

**2.** Frequently, an ODE is expressed in terms of the underlying function, for example, $\dfrac{dy}{dx} = y - 1$. Observe that $y = e^x + 1$ satisfies the given ODE, since $\dfrac{dy}{dx} = \dfrac{d}{dx}(e^x + 1) = e^x = y - 1$.

Consider the list of functions given in part 1. For each of the following ODEs, find the function in the list that satisfies the ODE.

| Differential Equation | Solution |
|---|---|
| $\dfrac{dy}{dx} = y - 1$ | $y = e^x + 1$ |
| $\dfrac{dy}{dx} = \dfrac{1}{y}$ | |
| $y' = -\dfrac{y}{x}$ | |
| $\dfrac{dy}{dx} = \dfrac{2y}{x}$ | |
| $y' = -\dfrac{x}{y}$ | |
| $\dfrac{dy}{dx} = y$ | |
| $\dfrac{dy}{dx} = -2y$ | |
| $\dfrac{dy}{dx} = m$ | |
| $y' = \dfrac{3y}{x}$ | |

**3.** ODEs can also be represented by verbal descriptions, such as "The derivative of the function is proportional to the function," which means that the derivative is a scalar multiple of the function. This statement is satisfied, for instance, by $y' = -0.5y$.

Consider the list of differential equations given in part 2. For each of the following statements, find an ODE in the list that satisfies the statement.

**(1)** The derivative of the function is proportional to the function.

**(2)** The rate of change of the function is constant.

**(3)** The derivative of the function equals the function.

**(4)** The rate of change of the function is inversely proportional to the function.

**(5)** The derivative of the function is one less than the function.

One of the most straightforward types of differential equations to solve is a *separable* ODE. In this case, the ODE can be represented by an equivalent equation which has the form

$$h(y) \, dy = g(x) \, dx.$$

Observe that all the *x*-terms appear on one side of the equation, whereas all the *y*-terms appear on the other side of the equation; that is, the variables can be "separated."

Separable ODEs can be solved by trying to find the antiderivative of each side of the equation. The solution is called an *implicit solution*, since it is an equation containing *x*'s and *y*'s, where *y* is assumed to be a function of *x*. For example, to use separation of variables to solve $\dfrac{dy}{dx} = -\dfrac{x}{y}$, you would do the following:

$$\frac{dy}{dx} = -\frac{x}{y} \qquad \text{Given ODE.}$$

$$y \, dy = -x \, dx \qquad \begin{array}{l}\text{Equivalent ODE with} \\ \text{variables separated.}\end{array}$$

$$\int y \, dy = \int (-x) \, dx \qquad \begin{array}{l}\text{Take the antiderivative} \\ \text{of both sides.}\end{array}$$

$$\tfrac{1}{2}y^2 = -\tfrac{1}{2}x^2 + C \qquad \text{Evaluate.}$$

Moreover, since *C* is any constant,

$$y^2 = -x^2 + C,$$

where the "new" *C* is twice the "old" *C*. This solution is the implicit solution to the given ODE. In this example, however, you can solve for *y*:

$$y(x) = \pm\sqrt{-x^2 + C},$$

where the notation $y(x)$ indicates that *y* is a function of *x*, and the domain of *y* depends on the value of *C*. In either case, the solution is said to be a *general solution*, since there are an infinite number of solutions—one for each value of the constant *C*.

One of the nice things about differential equations is that you can use differentiation to check your solution. For instance, you can use implicit differentiation to show that $y^2 = -x^2 + C$ is the solution:

$$y^2 = -x^2 + C \qquad \text{Implicit solution.}$$

$$2yy' = -2x \qquad \text{Differentiate implicitly.}$$

$$\frac{dy}{dx} = -\frac{x}{y} \qquad \text{Solve for } y' = \frac{dy}{dx}.$$

Or, you can differentiate $y(x) = \pm\sqrt{-x^2 + C}$ directly to show that it is the solution:

$$y(x) = \sqrt{-x^2 + C} \qquad \text{Explicit solution.}$$

$$y' = \tfrac{1}{2}(-x^2 + C)^{-1/2}(-2x) \qquad \text{Differentiate with respect to } x.$$

$$y' = -\frac{x}{\sqrt{-x^2 + C}} \qquad \text{Simplify.}$$

$$y' = -\frac{x}{y} \qquad \text{Substitute } y \text{ for } \sqrt{-x^2 + C}.$$

## Task 9-7: Solving Separable Equations

1. Find a general solution to each of the following ODEs, using the method of separation of variables. Check your solution, using implicit differentiation.

   **a.** $\dfrac{dy}{dx} = y$

   **b.** $y' = \dfrac{x^2}{y^2}$

   **c.** $x\,dy = y\,dx$

2. Since the general solution to a differential equation contains the parameter $C$, which can be any real number, the graphs of the general solution are *families of curves*. For each of the following ODEs (which you considered in part 1):

   **i.** Show that your solution is equivalent to the solution given below.

   **ii.** Use your CAS to plot the given solution for $C = 0$, $\pm 1$, and $\pm 2$ on the same pair of axes. Place your graph in your activity guide. Label each solution with its associated value of $C$.

**iii.** Describe the shape of the graphs of the members in the family of solutions, and describe how the value of $C$ affects the graph. Do you notice any generalities? Are the solutions bounded? Oscillatory? Increasing? Decreasing? Concave up? Concave down? Are there maxima? Minima?

**a.** $\dfrac{dy}{dx} = y$, where $y = Ce^x$.

**(1)** Show that your solution is equivalent to the given solution.

When you used separation of variables in part 2 to find the general solution to this ODE, you probably concluded that $\ln y = x + C$ or possibly that $y = e^{x+C}$ was the general solution, which was correct since your solution checked. However, $y = Ce^x$ is the way solutions to ODEs of this type are usually represented. How can you convert $\ln y = x + C$ to $y = Ce^x$? In other words, why are these two solutions equivalent? The answer involves the way the constant part of the solution is defined.

$\ln y = x + C$    General solution.

$y = e^{x+C}$    Convert to exponential form.

$y = e^x e^C$    Use the property that $e^{a+b} = e^a e^b$.

$y = Ce^x$    Note that $e^C$ is a constant. Replace the "old" constant $e^C$ by a "new" constant $C$ to arrive at the equivalent form of the general solution.

This is a standard trick. You will probably have to use a variation of it in part c below.

**(2)** Plot $y = Ce^x$ for $C = 0, \pm 1$, and $\pm 2$ on the same pair of axes. Place your graph below. Label.

**(3)** Describe the shape of the graphs of the members in the family of solutions, and describe how the value of $C$ affects the graph.

**b.** $y' = \dfrac{x^2}{y^2}$, where $y = \sqrt[3]{x^3 + C}$.

  **(1)** Show that your solution is equivalent to the given solution.

  **(2)** Plot $y = \sqrt[3]{x^3 + C}$ for $C = 0, \pm 1$, and $\pm 2$ on the same pair of axes. Place your graph below. Label.

  **(3)** Describe the shape of the graphs of the members in the family of solutions, and describe how the value of $C$ affects the graph.

**c.** $x\,dy = y\,dx$, where $y = Cx$ and $x \neq 0$.

  **(1)** Show that your solution is equivalent to the given solution.

  **(2)** Plot $y = Cx$ where $x \neq 0$, for $C = 0, \pm 1$, and $\pm 2$ on the same pair of axes. Place your graph below. Label.

**(3)** Describe the shape of the graphs of the members in the family of solutions, and describe how the value of $C$ affects the graph.

**d.** $\dfrac{dy}{dx} = -\dfrac{x}{y}$, where $y = \pm\sqrt{-x^2 + C}$.

*Note: This is the ODE that we discussed before you began Task 9-7.*

**(1)** The value of $C$ must be greater than or equal to 0. Explain why.

**(2)** Find the domain of $y$ in terms of $C$.

**(3)** Plot $y = \pm\sqrt{-x^2 + C}$ for $C = 0$, 1, 2, and 4 on the same pair of axes. Place your graph below. Label.

**(4)** Describe the shape of the graphs of the members in the family of solutions, and describe how the value of $C$ affects the graph.

**3.** In part 1, you used the method of separation of variables to find a general solution to a given ODE. Often, in applications, you need to find a particular solution. For instance, you might want to find the solution to

$\dfrac{dy}{dx} = -\dfrac{x}{y}$, where $y(1) = 3$. This is called an *initial value problem* or *IVP*. The solution to an IVP is the member of the family of solutions whose graph passes through the specified point, which in this case is $P(1,3)$.

To solve an IVP, you first find the general solution to the ODE, and then find the particular value of $C$ which makes the given initial condition hold. For example, you know that $y(x) = \pm\sqrt{-x^2 + C}$ is the general solution to $\dfrac{dy}{dx} = -\dfrac{x}{y}$. Thus, for $y(1) = 3$ to hold, you could solve for $C$ as follows:

$$y(x) = \pm\sqrt{-x^2 + C} \qquad \text{General solution.}$$
$$y(1) = \sqrt{-(1^2) + C} = 3 \quad \text{Let } y(1) = 3.$$
$$C = 10 \qquad\qquad\qquad \text{Solve for } C.$$

Thus, $y(x) = \sqrt{-x^2 + 10}$ is the solution to this IVP. Note that we chose the positive square root in the general solution, instead of the negative square root. Why?

Solve the following IVPs.

**a.** $(1 + x)y' + (1 + y) = 0$, $y(2) = 6$

**b.** $x^2 y' - \dfrac{x^2}{y} = \dfrac{1}{y}$, $y(1) = 5$

Up to this point, you have been able to find a solution to a given initial value problem by separating the variables and finding the antiderivative of each side of the equation. The antiderivative always exists, but sometimes you may not be able to find a *closed form* for an antiderivative—for example, you cannot find a closed form for $\int e^{x^2}$ (see Task 8-12). In cases such as this, you can use numerical methods to approximate the solution to an IVP, just as you used numerical methods to approximate a definite integral.

For instance, given the rate of change of a function and an initial value, you can use *Euler's Method* to approximate the graph of the solution, even though you do not know a formula for the function. The general idea is as follows: Starting at the given initial value, approximate the next value by "walking" along the tangent line whose slope is determined by the given rate of change. Repeat the process; that is, starting at the current value, walk along the tangent line to approximate the next value, and so on. Euler's Method gives a very crude approximation of the exact solution. Other methods give much better approximations, but with Euler's Method, it is easy to visualize what is happening, so it is a good method with which to start.

The next task guides you through the process of applying Euler's Method. In the first part of the task, you will consider the ideas underlying the method—in particular, how to "walk" along a tangent line from one approximate value to the next. Next you will develop the general formula for the method. Finally, you will improve the approximation which you calculated in the first part by applying Euler's formula with a smaller step size. Although you usually approximate a solution to an IVP only when you are unable to find the solution directly, you will consider an IVP which you can solve. This will enable you to compare your approximate solution with the exact solution.

## Task 9-8: Utilizing Euler's Method

1. Consider the initial value problem

$$\frac{dy}{dt} = t,\ y(0) = 1.$$

In the following parts, we will guide you through how to use Euler's Method to approximate some values of the solution to the IVP over the closed interval [0,6], where $\Delta t = 2$. You will use these points to sketch an approximate graph of the solution.

Note that $\Delta t$ is called the *step size* or *time step*. It determines how many values will be calculated. In this case, since you are interested in the interval [0,6], where $\Delta t = 2$, you will estimate values of the solution $y$ when $t = 2$, 4, and 6. Usually the more points you consider, the more accurate your approximation is.

a. The initial value, which is given in the statement of the IVP, provides you with the first point $P(t_0, y_0)$ on the graph of the approximate solution.

In this case, $y(0) = 1$, so $t_0 = 0$ and $y_0 = 1$. Plot $P(0,1)$ on the axes given below.

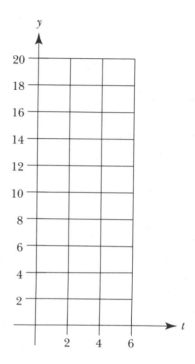

**b.** To approximate the solution at the next value of $t$, which is denoted by $t_1$, find the slope of the tangent line to the graph of the solution at $P(t_0, y_0)$, and then move along the tangent line until it intersects the vertical line $t = t_1$. The $y$-coordinate of the point of intersection, $P(t_1, y_1)$, provides the next approximation to the solution.

Note that $t_1 = t_0 + \Delta t$. So, in this case, $t_1 = 0 + 2 = 2$.

**(1)** Use the fact that $\dfrac{dy}{dt} = t$ to find the slope of the tangent line to the graph of the solution at $P(t_0, y_0)$—that is, at $P(0,1)$.

**(2)** Starting at the point $P(0,1)$, move along the tangent line until it intersects the vertical line $t_1 = 2$.

  **(a)** Mark the point of intersection $P(t_1, y_1)$ on the set of axes in part a.

  **(b)** State the value of $y_1$.

**c.** Approximate the solution at the next value of $t$, $t_2 = t_1 + \Delta t$, by repeating the process you used in part b where, in this case, $t_2 = 2 + 2 = 4$.

**(1)** In part b, you should have determined that $P(t_1, y_1) = P(2,1)$. If you did not, backtrack and check your calculations.

Find the slope of the tangent line to the graph of the solution at the last estimate $P(t_1, y_1)$—that is, at $P(2,1)$.

**(2)** Starting at the point $P(2,1)$, move along the tangent line until it intersects the vertical line $t_2 = 4$.

    **(a)** Mark the point of intersection $P(t_2,y_2)$ on the set of axes in part a.

    **(b)** State the value of $y_2$.

**d.** Last time. Repeat the process to approximate the solution at the next value of $t$, $t_3 = t_2 + \Delta t$, where, in this case, $t_3 = 6$.

    **(1)** In part c, you should have determined that $P(t_2,y_2) = P(4,5)$. Find the slope of the tangent line to the graph of the solution at the last estimate, $P(t_2,y_2)$.

    **(2)** Starting at the point $P(4,5)$, move along the tangent line until it intersects the vertical line $t_3 = 6$.

        **(a)** Mark the point of intersection $P(t_3,y_3)$ on the set of axes in part a.

        **(b)** State the value of $y_3$.

**e.** On the set of axes in part a, sketch the graph of the approximate solution by connecting the four points that you plotted in parts a through d.

**f.** Summarize your results from parts a through d by filling in the following table, where $y_i'$ is the value of the slope of the tangent line at $P(t_i,y_i)$, where $i = 0, 1, 2, 3$. Observe that we have completed the first line, which lists the coordinates of the initial point $P(t_0,y_0)$ and the slope of the tangent line at that point $y_0'$.

| $i$ | $t_i$ | $y_i$ | $y_i'$ |
|---|---|---|---|
| 0 | 0 | 1 | 0 |
| 1 |  |  |  |
| 2 |  |  |  |
| 3 |  |  |  |

**g.** Compare the approximate solution, which you found using Euler's Method with $\Delta t = 2$, to the exact solution.

    **(1)** Use the method of separation of variables to find the exact solution to the given initial value problem $\dfrac{dy}{dt} = t$, where $y(0) = 1$.

(2) Sketch the graph of the exact solution, for $0 \le t \le 6$, on the axes in part 1a.

(3) Describe the relationship between the graph of the exact solution and the graph of the approximate solutions as $t$ moves away from the initial value $t_0 = 0$.

(4) Calculate the size of the error for each of the approximate values, and enter your data in the following table.

| $i$ | $t_i$ | Approximate solution $y_i$ | Exact solution $y$ | Error $y - y_i$ |
|-----|-------|---------------------------|--------------------|-----------------|
| 0   | 0     |                           |                    |                 |
| 1   | 2     |                           |                    |                 |
| 2   | 4     |                           |                    |                 |
| 3   | 6     |                           |                    |                 |

(5) Describe how you might decrease the size of the error. In other words, what might you do to improve the accuracy of your approximate solution?

2. In this part, you will develop the general formula for using Euler's method to approximate a solution to a given IVP.

First, some notation. Let

$$\frac{dy}{dt} = F(t,y), \quad \text{where } y(t_0) = y_0$$

be the given IVP. For $1 \le i \le n$, let

$$t_i = t_{i-1} + \Delta t$$

be $n$ equally distributed points, over a given interval $[a,b]$. You will be given either the value of $n$ or the step size $\Delta t$ between successive approximations. In either case, observe that

$$\Delta t = (b - a)/n.$$

Let

$y_i$ be the approximate solution at $t = t_i$.

The goal is to find a formula for $y_i$, where $y_i$ is determined by using information about the approximation at the previous value of $t$, $t_{i-1}$. In particular, $y_i$ is determined as follows:

- Use the given rate of change to calculate the slope of the tangent line at the previous point, $P(t_{i-1}, y_{i-1})$.
- Walk along the tangent line until it intersects the vertical line $t = t_i$.
- Define $y_i$ to be the second coordinate of the point of intersection.

**a.** Find a formula for $y_1$.

Note that $P(t_0, y_0)$ is given in the statement of the IVP. So, you can immediately plot this point. Moreover, you can find the value of the slope of the tangent line to the graph of the solution at $P(t_0, y_0)$ by evaluating the formula for $\dfrac{dy}{dt}$ (which is also given in the statement of the IVP) at $t = t_0$ and $y = y_0$. Let $y_0'$ be the value of the slope at this point—that is, suppose

$$y_0' = \frac{dy}{dt}\bigg|_{P(t_0, y_0)}.$$

Consider the following diagram where the value of $y_1$ is unknown.

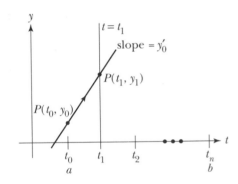

**(1)** Note that $\Delta t = t_1 - t_0$. Find the formula for $y_0'$ in terms of $y_0$, $y_1$, and $\Delta t$.

$$y_0' = \underline{\hspace{3cm}}.$$

**(2)** Solve for $y_1$. Your formula should be in terms of $y_0'$, $y_0$, and $\Delta t$, all of whose values are known.

$$y_1 = \underline{\hspace{3cm}}.$$

**b.** Now that you know the value of $y_1$, you can calculate the slope of the tangent line at $P(t_1, y_1)$ by evaluating

$$y_1' = \frac{dy}{dt}\bigg|_{P(t_1, y_1)}$$

and then repeat the process described in part a to find $y_2$. You can then use the value of $y_2$ to find $y_3$, and so on, up to $y_{i-1}$. Let $y_{i-1}'$ be the value of the slope of the tangent line at the last known point—that is, suppose

$$y_{i-1}' = \frac{dy}{dt}\bigg|_{P(t_{i-1}, y_{i-1})}.$$

Consider the following diagram where the value of $y_i$ is unknown.

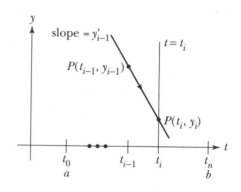

**(1)** Note that $\Delta t = t_i - t_{i-1}$. Find the formula for $y'_{i-1}$ in terms of $y_{i-1}$, $y_i$, and $\Delta t$.

$$y'_{i-1} = \underline{\hspace{4cm}}.$$

**(2)** Solve for $y_i$. Your formula should be in terms of $y'_{i-1}$, $y_{i-1}$, and $\Delta t$, all of whose values are known.

$$y_i = \underline{\hspace{4cm}}.$$

**3.** In the last part, you derived *Euler's formula*:

$$y_1 = y_0 + y'_0 \Delta t$$
$$y_2 = y_1 + y'_1 \Delta t$$
$$y_3 = y_2 + y'_2 \Delta t$$
$$\vdots$$

or

$$y_i = y_{i-1} + y'_{i-1}\Delta t \quad \text{for } 1 \le i \le n,$$

where $P(t_0, y_0)$ is known and $\Delta t$ is given. Use this formula to approximate the solution to the IVP which you considered in part 1, namely,

$$\frac{dy}{dt} = t, \quad y(0) = 1.$$

This time, suppose $\Delta t = 0.5$.

**a.** Complete the following table.

*Note: This is a good opportunity to use a spreadsheet package, the table feature on your graphing calculator, or your CAS to do your calculations.*

| $i$ | $t_i$ | $y_i$ | $y_i'$ |
|---|---|---|---|
| 0 | 0 | 1 | |
| 1 | 0.5 | | |
| 2 | 1 | | |
| 3 | 1.5 | | |
| 4 | | | |
| 5 | | | |
| 6 | | | |
| 7 | | | |
| 8 | | | |
| 9 | | | |
| 10 | | | |
| 11 | | | |
| 12 | | | |

**b.** On the set of axes in part 1, sketch the graph of this approximate solution by plotting and connecting the 13 points in the above table.

**c.** How does this approximation compare the first approximation which you derived in part 1? How does it compare to the exact solution?

## Unit 9 Homework After Section 2

- Complete the tasks in Section 2. Be prepared to discuss them in class.

- Solve some related rates problems in HW9.6.

**HW9.6** Solve the following related rates problems.*

1. Two cars start from the same point at the same time. One car travels east at 60 mph and the other travels north at 25 mph. How fast is the distance between them (as the crow flies) increasing at the end of 1 hour?

2. The volume of a tumor is given by $V = \frac{4}{3}\pi r^3$. The radius is increasing by 0.03 centimeters per day (cm/day) at the moment when $r = 1.2$ cm. How fast is the volume changing at that moment?

3. The area of a healing wound is given by $A = \pi r^2$. The radius is decreasing at the rate of 1 millimeter per day (mm/day) at the moment when $r = 25$ mm. How fast is the area decreasing at that moment?

4. According to Poiseuille's Law, the flow of blood in a blood vessel is faster toward the center of the vessel and slower toward the outside. The speed of blood $V$ is given by $V = \frac{p}{4L\nu}(R^2 - r^2)$, where $R$ is the radius of the blood vessel, $r$ is the distance of the blood from the center of the vessel, and $p$, $L$, and $\nu$ are physical constants related to pressure, length, and viscosity of the blood vessels, respectively. Use this formula to solve the following problem:

   When shoveling snow in cold air, a person with a history of heart trouble can develop angina (chest pains) due to contracting blood vessels. To counteract this, he or she may take nitroglycerin tablets, which dilates the blood vessels. Suppose that after a nitroglycerin tablet is taken, the radius of a blood vessel, $R$, dilates at the rate of 0.0025 millimeters per minute (mm/min) at the moment when $R = 0.02$ mm. Assume that $r$ is a constant, as well as $p$, $L$, and $\nu$. What is the rate of change of the speed of the blood at that moment?

5. A point travels along the graph of $x^2 - y^2 = 9$, where $\frac{dx}{dt} = \frac{1}{x}$. Find $\frac{dy}{dt}$ at the point $P(5,4)$.

6. Gas is escaping from a spherical balloon at the rate of 10 ft³/hr. At what rate is the radius changing when the volume is 400 ft³?

---

*See *Applied Calculus*, 3rd ed., by Bittenger and Morrel, Addison-Wesley, Reading, MA, 1993.

7. Suppose a spherical snowball is melting and the radius is decreasing at a constant rate, changing from 12" to 8" in 45 minutes. How fast was the volume changing when the radius was 10"?

8. A softball diamond has the shape of a square with sides 60 feet long. If a player is running from second base to third base at a speed of 24 feet per second (ft/sec), at what rate is her distance from home plate changing when she is 20 feet from third?

9. A stone is dropped into a lake, causing circular ripples whose radii increase at a constant rate of 0.5 meters per second (m/sec). At what rate is the circumference of the ripple changing when its radius is 4 meters?

10. A boy flying a kite lets out the string at a rate of 2 feet per second (ft/sec). The kite moves horizontally at an altitude of 100 feet. Assuming there is no sag in the string, find the rate at which the kite is moving when he has let out 125 feet of string.

- Solve some ODEs and IVPs in HW9.7.

### HW9.7

1. Show that $y$ is a solution to the given ODE, in each of the following exercises.

   a. Show that $y = Ae^x \cos(x) + Be^x \sin(x)$ is a general solution to $y'' - 2y' + 2y = 0$.

   b. Show that $y = \dfrac{C}{x^{2/3}}$ is a general solution to $2xy^3\, dx = -3x^2y^2\, dy$.

   c. Show that $x^2 - y^2 = C$ is a general solution to $y\,y' = x$.

   d. Show that $y = Cx^3$ is a general solution to $\dfrac{dy}{dx} = \dfrac{3y}{x}$.

   e. Show that $y = C_1e^x + C_2e^{2x}$ is a general solution to $3\dfrac{dy}{dx} - \dfrac{d^2y}{dx^2} = 2y$.

2. Use the method of separation of variables to find a general solution to each of the following differential equations. Check your solution.

   a. $\dfrac{dy}{dx} = e^{-2x} + 2x$

   b. $\cos(x)\, dx - \sin(y)\, dy = 0$

   c. $y'(x) = 2y(x)$

   d. $\dfrac{dy}{dx} = e^{x-y}$

   e. $x^2y' - yx^2 = y$

   f. $2xy + 2x = (x^2 + 1)\dfrac{dy}{dx}$

3. Solve the following IVPs.

   a. $y' = -5y,\ y(0) = 10$

   b. $x^2\, dy - \tfrac{1}{2}y^3\, dx = 0,\ y(-1) = 1$

   c. $2x\dfrac{dy}{dx} = 3y,\ y = 4$ when $x = 2$

   d. $y'' = 3 - e^x,\ y(0) = 1$ and $y'(0) = 7$

- Examine an "uninhibited growth model" in HW9.8.

**HW9.8** A function $y = y(t)$ and its derivative $\dfrac{dy}{dt}$ are *proportional* whenever $\dfrac{dy}{dt} = ky$ for some constant $k$. In this case, the slope of the tangent line to the graph of the function at any point equals $k$ times the value of the function at that point. When $k > 0$, $\dfrac{dy}{dt} = ky$ is the basic model for *uninhibited population growth*.

1. Show that $\dfrac{dy}{dt} = ky$ if and only if $y = Ce^{kt}$ as you complete the following steps.

   a. Assume $\dfrac{dy}{dt} = ky$. Show that $y = Ce^{kt}$.

   b. Assume that $y = Ce^{kt}$. Show that $\dfrac{dy}{dt} = ky$.

   c. In this case, the value of $C$ can be determined by evaluating $y$ at $t = 0$. Show that $C = y(0)$.

2. Suppose the SuperDuper Savings Bank compounds interest continuously at a rate of 20% per year. Thus, a balance $P$ grows at the rate of $\dfrac{dP}{dt} = 0.20\,P$.

   a. Find $P$. What does $C$ represent?

   b. Suppose at age 10 you spent your summer doing odd jobs around your neighborhood. You saved all the money you earned, and at the end of the summer you deposited $500 at the SuperDuper.

      (1) What was the balance in your account at the end of 1 year?

      (2) How long did it take for your money to double?

      *Hint: Use natural logarithms to find the answer.*

      (3) You promised your parents that when you were 18, you would contribute $2500 toward the first year of your college tuition. Assuming your initial $500 was the only money you deposited, was there enough money in your account at the SuperDuper to cover your promise?

3. You take over a franchise pizza business, which consists of 20 franchises. Your goal as the CEO is to increase the number of franchises at a rate of 15% per year. Assuming you manage to meet this goal:

   a. How many franchises will you own in 10 years?

   b. How many years will it take for you to double your initial number of franchises?

   c. If you purchase the business at age 26 and plan to retire when you own 100 franchises, how old will you be when you retire?

**4.** Your hometown bank claims that your money will double in 10 years. Find the bank's annual interest rate, assuming the bank compounds interest continuously.

- Use Euler's Method to approximate solutions to IVPs in HW9.9.

**HW9.9**

**1.** Consider $\dfrac{dy}{dx} = \dfrac{2}{x}$, where $y(1) = 0$.

    **a.** Use Euler's Method with $\Delta x = 0.5$ to approximate $y(5)$.

    **b.** Sketch a graph of an approximate solution to the given IVP for $1 \le x \le 5$ by plotting and connecting the points which you calculated in part a.

**2.** Consider $y' = y$, where $y(0) = 1$.

    **a.** Use Euler's Method with $n = 5$ to make a careful sketch of the graph of an approximate solution on the interval $[0,1]$.

    **b.** Since $y = e^x$ is the exact solution to this IVP, $y_5 \approx y(1) = e$. How close is your approximation to $e$?

**3.** Consider $\dfrac{dy}{dt} = \dfrac{t}{y}$, where $y(2) = -1$.

    **a.** Find the exact solution.

    **b.** For $2 \le t \le 2.5$, compare the values of the exact solution to the approximate solution found using Euler's Method with $\Delta t = 0.1$. Use your CAS or calculator to compute the values to five decimal places. List your findings in a table.

    **c.** Are the Euler approximations too large or too small when $\Delta t = 0.1$? Will this always be the case, regardless of the size of $\Delta t$? That is, will the Euler approximations always be too large (too small)? Justify your response.

**4.** Suppose the rate of growth of yeast cells in the lab is $\dfrac{dY}{dt} = 0.008\,Y(700 - Y)$, where $Y(t)$ is the number of yeast cells after $t$ hours. Suppose you initially have 10 cells.

    **a.** Use Euler's Method with $\Delta t = 0.5$ to approximate the numbers of yeast cells after 2 hours.

    **b.** Use Euler's Method with $\Delta t = 1$ to approximate $Y(6)$.

- Numerical methods provide ways to estimate solutions. For instance, you can use the Trapezoidal Rule to approximate a definite integral and Euler's Method to find an approximate solution to an initial value problem. Newton's Method provides a way to approximate a real root of a function. Consider the idea underlying Newton's Method and use the approach to solve some problems in HW9.10.

**HW9.10** To use Newton's Method to approximate a real root to an equation $f(x) = 0$, first choose an initial guess or an initial approximation, say $x_0$, for the root, where $f'(x_0) \neq 0$. Then, use $x_0$ to find the next guess, say $x_1$, as follows:

- Construct the tangent line to the graph of $f$ at $P(x_0, f(x_0))$.

- Find the place where the tangent line intersects the horizontal axis. Call this $x_1$.

Repeat the process until the desired number of digits are repeated from one approximation to the next. For example, if you wanted to approximate a solution to three decimal places and you found that $x_4 = 1.453\,21$, $x_5 = 1.679\,01$, and $x_6 = 1.679\,25$, you would terminate the process and conclude that $f(x) = 0$ when $x \approx 1.679$.

1. Derive the general formula for Newton's Method. Consider the following graph:

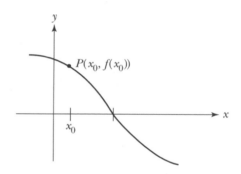

    **a.** Find a formula for $x_1$.

        **(1)** Use a straightedge to sketch the tangent line to the graph of $f$ at $P(x_0, f(x_0))$.

        **(2)** Assume the tangent line intersects the horizontal axis at $x = x_1$. Label it $x_1$.

        **(3)** Note that the tangent line passes through the points $P(x_0, f(x_0))$ and $P(x_1, 0)$, where the slope of the line is $f'(x_0)$. Find a formula for the slope in terms of $x_0$, $x_1$, and $f(x_0)$.

        **(4)** Solve your equation in part (3) for $x_1$. Your formula for $x_1$ should be in terms of $x_0$, $f(x_0)$ and $f'(x_0)$, all of which have known values.

    **b.** Find a formula for $x_2$ in terms of $x_1$, $f(x_1)$, and $f'(x_1)$ by repeating the process described in part a. Label the graph appropriately.

    **c.** Generalize your observations. Give the general formula for $x_i$ in terms of $x_{i-1}$, $f(x_{i-1})$, and $f'(x_{i-1})$, where $i = 1, 2, 3, \ldots$.

2. In part 1, you should have discovered that $x_i = x_{i-1} - \dfrac{f(x_{i-1})}{f'(x_{i-1})}$, for $i = 1, 2, 3, \ldots$, provided $f'(x_{i-1}) \neq 0$, where $x_0$ is given. This technique of successive approximations of real roots is called *Newton's Method*.

   a. Use Newton's Method to approximate $\sqrt{7}$ to three decimal places. *Hint*: Consider $f(x) = x^2 - 7$ with $x_0 = 3$.

   b. Use Newton's Method to find the point of intersection of the graphs of $y = e^{-x}$ and $y = x$ to five decimal places. Sketch the graphs and label the point of intersection. *Hint*: Consider $f(x) = x - e^{-x}$ with $x_0 = 0.25$.

   c. Use Newton's Method to find all the points of intersection of $y = \cos(x)$ and $y = x^2$.

   d. Use Newton's Method to find all the roots of $x^4 + 2x^3 - 5x^2 + 1 = 0$ which lie between 1 and 2.

   e. Use Newton's Method to find all the roots of $x^3 - 3x + 1 = 0$.

3. As you may have discovered in part 2, Newton's Method does not always generate a sequence $x_0, x_1, x_2, \ldots$ that converges to a solution of $f(x) = 0$. Explain why the method does not work in each of the following situations.

   a. Newton's Method does not apply if $f'(x_n) = 0$ for some value of $n$.

      **(1)** Sketch a graph of $f$ near $P(x_n, f(x_n))$, where $f'(x_n) = 0$.

      **(2)** Explain why it is impossible to locate $x_{n+1}$ on the graph in this case.

      **(3)** Explain why it is impossible to use Newton's formula to calculate $x_{n+1}$ in this case.

   b. Newton's Method does not apply if $f$ has no real roots. Consider, for example, $f(x) = x^2 + 1$ with $x_0 = 1$. Describe what happens in this case.

   c. If $x_0$ is not chosen sufficiently close to a root, the sequence generated by Newton's Method, $x_0, x_1, x_2, \ldots$, may diverge.

      **(1)** For instance, consider the following graph:

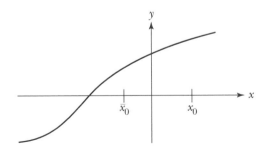

      Explain why it appears that $x_0, x_1, x_2, \ldots$ does not converge to the root, whereas it appears that $\tilde{x}_0, \tilde{x}_1, \tilde{x}_2, \tilde{x}_3, \ldots$ does.

**(2)** It is important to choose $x_0$ "wisely." One aspect of making a wise choice is to choose $x_0$ near a root. It is also important to stay away from places where $f$ has a local extremum or where the graph levels off. Explain why this is the case.

## SECTION 3

### More on Integration

The Fundamental Theorem of Calculus says that if $f$ is continuous on the closed interval $[a, b]$, then

$$\int_a^b f(x)\ dx = F(b) - F(a),$$

where $F$ is any antiderivative of $f$. Moreover, $\int_a^b f(x)\ dx$ can be associated with the area of the region bounded by the graph of $f$, where $a \le x \le b$.

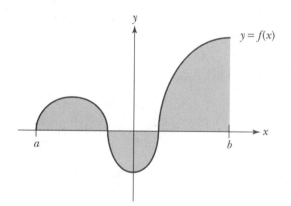

In the next task, you will investigate integrals where:

• One or both of the limits of integration is infinite, for example,

$$\int_{-\infty}^0 e^x\ dx \quad \text{or} \quad \int_1^\infty \frac{1}{x^2}\ dx \quad \text{or} \quad \int_{-\infty}^\infty e^{-x^2}\ dx.$$

• The function is unbounded, for example,

$$\int_1^4 \frac{1}{(x-1)^2}\ dx.$$

Integrals of this type are called *improper integrals.* Observe that the region associated with an improper integral is not bounded. It is possible, however, for an improper integral to have a *finite* value, even though the associated region is of *infinite* extent. You will investigate this possibility in the next task.

## Task 9-9: Evaluating Improper Integrals

1. Examine some improper integrals where the upper and/or the lower bound of integration is infinite.

   **a.** Consider $\int_1^\infty \frac{1}{x^2}\, dx.$

   **(1)** Use your CAS to graph $f(x) = 1/x^2$. Place a copy of the graph below. Shade the region associated with the improper integral, noting that the region is not bounded since the upper limit of integration is infinite.

   **(2)** You cannot evaluate $\int_1^\infty \frac{1}{x^2}\, dx$ directly, since it makes no sense to try to evaluate an antiderivative of $f(x) = 1/x^2$ at infinity. However, you can evaluate $\int_1^b \frac{1}{x^2}\, dx$, where $b$ is an arbitrarily large number.

   Make a conjecture about the value of $\int_1^\infty \frac{1}{x^2}\, dx$ by considering the limiting behavior of $\int_1^b \frac{1}{x^2}\, dx$ as $b \to \infty$.

   **(a)** Use your CAS to evaluate $\int_1^b \frac{1}{x^2}\, dx$ as $b$ gets large. Record your results in the following table.

| $b$ | 10 | 100 | 1,000 | 10,000 | 100,000 |
|---|---|---|---|---|---|
| $\int_1^b \frac{1}{x^2}\, dx$ | | | | | |

   **(b)** Based on your calculations, it appears that

   $$\int_1^b \frac{1}{x^2}\, dx \to \boxed{\phantom{x}}, \quad \text{as } b \to \infty.$$

   **(3)** In general, if $f$ is continuous on the interval $[a, \infty)$, then

   $$\int_a^\infty f(x)\, dx = \lim_{b \to \infty} \int_a^b f(x)\, dx,$$

provided the limit exists. If it does exist, the improper integral $\int_a^\infty f(x)\,dx$ is said to *converge*; otherwise, it is said to *diverge*.

In part (2), you probably conjectured that $\lim\limits_{b\to\infty} \int_1^b \dfrac{1}{x^2}\,dx = 1$. Use calculus to show that this conjecture is true, by filling in the steps corresponding to the explanations given below.

$$\int_1^\infty \frac{1}{x^2}\,dx = \lim_{b\to\infty}\int_1^b \frac{1}{x^2}\,dx \qquad \text{Definition of improper integral.}$$

$$= \qquad\qquad \text{Use the Fundamental Theorem of Calculus to evaluate } \int_1^b \frac{1}{x^2}\,dx.$$

$$= \qquad\qquad \text{Evaluate the limit, noting that } \frac{1}{b}\to 0, \text{ as } b\to\infty.$$

**b.** Not all improper integrals converge. For example, even though the graphs of $f(x) = 1/x^2$ and $f(x) = 1/x$ have the same general shape for $x \geq 0$, $\int_1^\infty \dfrac{1}{x^2}\,dx$ converges, while $\int_1^\infty \dfrac{1}{x}\,dx$ diverges. Show that this is the case.

**(1)** Graph $f(x) = 1/x$. Shade the region associated with the improper integral.

**(2)** Use your CAS to show that it is reasonable to conjecture that $\lim\limits_{b\to\infty} \int_1^b \dfrac{1}{x}\,dx$ does not exist.

| $b$ | 10 | 100 | 1,000 | 10,000 | 100,000 |
|---|---|---|---|---|---|
| $\int_1^b \dfrac{1}{x}\,dx$ | | | | | |

**(3)** Show that $\int_1^\infty \dfrac{1}{x}\,dx$ diverges by evaluating $\int_1^b \dfrac{1}{x}\,dx$ and then taking the limit as $b\to\infty$. In this case, the limit will not be finite.

**c.** In the last two parts, you considered the case for which the upper limit of integration is infinite. Similarly, if $f$ is continuous on $(-\infty, b]$, then

$$\int_{-\infty}^{b} f(x)\ dx = \lim_{a \to -\infty} \int_{a}^{b} f(x)\ dx,$$

provided the limit exists. And, finally, if $f$ is continuous on $(-\infty, \infty)$, then

$$\int_{-\infty}^{\infty} f(x)\ dx = \int_{-\infty}^{c} f(x)\ dx + \int_{c}^{\infty} f(x)\ dx,$$

where $c$ is a real number, provided both the improper integrals on the right side of the equation converge.

For example, consider $\int_{-\infty}^{\infty} xe^{-x^2}\ dx$. Show that $\int_{-\infty}^{\infty} xe^{-x^2}\ dx = 0$, where $c = 0$.

**(1)** Graph $f(x) = xe^{-x^2}$. Shade the region associated with the improper integral. Explain why it is reasonable to conjecture that $\int_{-\infty}^{\infty} xe^{-x^2}\ dx$ converges to 0.

**(2)** Suppose $c = 0$. Then by definition,

$$\int_{-\infty}^{\infty} xe^{-x^2}\ dx = \int_{-\infty}^{0} xe^{-x^2}\ dx + \int_{0}^{\infty} xe^{-x^2}\ dx.$$

Use calculus to evaluate the two improper integrals on the right-hand side.

**(a)** Show that $\int_{-\infty}^{0} xe^{-x^2}\ dx = -\frac{1}{2}$.

**(b)** Similarly, show that $\int_0^\infty xe^{-x^2}\,dx = \frac{1}{2}$.

**(c)** Show that $\int_{-\infty}^\infty xe^{-x^2}\,dx = 0$ by combining your results from parts (a) and (b).

2. An integral $\int_a^b f(x)\,dx$ is also said to be improper if $f$ is continuous on the interval $[a,b]$, except that it is unbounded—that is, if the function has a blowup discontinuity and hence its graph has a vertical asymptote—at either the upper and/or the lower bound of integration.

For example, consider $\int_0^9 \frac{1}{\sqrt{x}}\,dx$.

**a.** Use your CAS to graph $f(x) = 1/\sqrt{x}$. Place a copy of the graph below. Shade the region associated with the improper integral, noting that the region is not bounded since the graph of $f$ has a vertical asymptote at $x = 0$.

**b.** You cannot use the Fundamental Theorem of Calculus to evaluate $\int_0^9 \frac{1}{\sqrt{x}}\,dx$, since $f(x) = 1/\sqrt{x}$ is not continuous at 0. However, you can evaluate $\int_c^9 \frac{1}{\sqrt{x}}\,dx$, where $c$ is any positive number close to 0. Make a conjecture about the value of $\int_0^9 \frac{1}{\sqrt{x}}\,dx$ by considering the limiting behavior of $\int_c^9 \frac{1}{\sqrt{x}}\,dx$ as $c$ approaches 0 from the right.

(1) Use your CAS to evaluate $\int_c^9 \frac{1}{\sqrt{x}}\, dx$ as $c \to 0^+$. Record your results in the following table.

| $c$ | 0.1 | 0.01 | 0.001 | 0.0001 | 0.00001 |
|---|---|---|---|---|---|
| $\int_c^9 \frac{1}{\sqrt{x}}\, dx$ | | | | | |

(2) Based on your calculations, it appears that

$$\int_c^9 \frac{1}{\sqrt{x}}\, dx \to \boxed{\phantom{xx}}, \quad \text{as } c \to 0^+.$$

c. In general, if $f$ is continuous on the interval $(a,b]$ and has a blowup discontinuity at $x = a$, then

$$\int_a^b f(x)\, dx = \lim_{c \to a^+} \int_c^b f(x)\, dx,$$

provided the limit exists. Similarly, if $f$ is continuous on the interval $[a,b)$ and has a blowup discontinuity at $x = b$, then

$$\int_a^b f(x)\, dx = \lim_{c \to b^-} \int_a^c f(x)\, dx,$$

provided the limit exists. In either case, if the limit exists, then the improper integral $\int_a^b f(x)\, dx$ is said to converge; otherwise, it is said to diverge.

In part (2), you probably conjectured that $\lim_{c \to 0^+} \int_c^9 \frac{1}{\sqrt{x}}\, dx = 6$. Show that this conjecture is true, by evaluating $\int_c^9 \frac{1}{\sqrt{x}}\, dx$ and then taking the limit as $c \to 0^+$.

Not only can you use integration to find the area of an unbounded region, you can also use it to find the volume of a solid of revolution. To generate a solid, you revolve a region in the $xy$-plane bounded by, say, $x = a$, $x = b$, the $x$-axis, and the graph of a function $f$, about a line, such as the $x$-axis. The line about which the region is revolved is called the *axis of revolution*. The resulting solid is called a *solid of revolution*. In the next task, you will construct some solids and find the formula for calculating the volume of a solid of revolution.

### Task 9-10: Finding the Volume of a Solid of Revolution Using the Disc Approach

1. Describe some solids of revolution. For each of the regions in parts a through e:

   **i.** Sketch the region in the $xy$-plane that generates the solid.

   **ii.** Sketch the solid that results when the region is revolved about the specified axis of revolution.

   **iii.** Describe the shape of the solid.

   **a.** The region in the first quadrant bounded by $x = 2$, $x = 5.5$, and $y = 3.25$. The axis of revolution is the $x$-axis.

   **b.** The region bounded by the semicircle located in the first and second quadrants that has radius 3 and center at the origin. The axis of revolution is the $x$-axis.

**c.** The region bounded by $y = x^2 + 1$ for $-2 \le x \le 2$ and the *x*-axis. The axis of revolution is the *x*-axis.

**d.** The region bounded by $f(x) = \sin(x)$ for $-\pi \le x \le \pi$ and the *x*-axis. The axis of revolution is the *x*-axis.

**e.** Let $a$ and $b$ be real numbers, where $a < b$. Let $f$ be a smooth curve, defined on $[a,b]$, and such that $f(x) \ge 0$ for $a \le x \le b$. Consider the region bounded by $f$ and the *x*-axis. The axis of revolution is the *x*-axis.

**2.** Now that you have a feel for what a solid of revolution looks like, construct some solids out of paper using scissors and a compass.

Construct the solid formed by revolving a function $f$, where $a \le x \le b$, about the *x*-axis. Carefully follow the instructions on the modeling template on page 351. Divide the problems among the people in your group, so that each member constructs a different solid. Use your calculator or CAS to evaluate $f$ at $x_0, x_1, \ldots, x_n$, where $x_0 = a$ and $x_n = b$.

**a.** $f(x) = \sqrt{x}$ for $1 \le x \le 9$, $n = 8$

**b.** $f(x) = \sin(x)$ for $0 \le x \le \pi$, $n = 4$

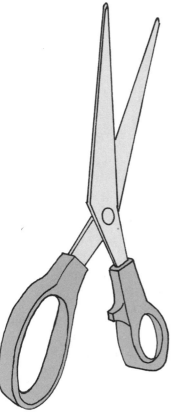

**c.** $f(x) = \dfrac{1}{x}$ for $1 \le x \le 8$, $n = 7$

**d.** $f(x) = x + 1$ for $2 \le x \le 9$, $n = 7$

3. Each of the circles that you cut out in part 2 represents the base of a slice of the solid of revolution determined by the graph of the given function. You can approximate the volume of the solid by calculating the volume of each slice, or disc, and finding the sum of the volumes. Observe that each disc is a *right circular cylinder*, since it is generated by revolving a rectangle about the *x*-axis, where

$$\text{Volume of disc} = (\text{area of the base}) \cdot (\text{height})$$

$$= \pi(\text{radius of disc})^2 \cdot (\text{thickness of disc})$$

$$= \pi r^2 h.$$

Use this approach to approximate the value of the solid of revolution that you constructed in the last part. Use your calculator or CAS to do calculations. Write the results in your activity guide. As before, divide the exercises among the members of your group.

**a.** $f(x) = \sqrt{x}$ for $1 \le x \le 9$, $n = 8$

**b.** $f(x) = \sin(x)$ for $0 \le x \le \pi$, $n = 4$

**c.** $f(x) = \dfrac{1}{x}$ for $1 \le x \le 8$, $n = 7$

**d.** $f(x) = x + 1$ for $2 \le x \le 9$, $n = 7$

4. Develop a general formula for approximating the volume of a solid of revolution using discs (right circular cylinders).

   **a.** In the space below, sketch an arbitrary solid of revolution generated by revolving a region bounded by the *x*-axis, the graph of *f*, $x = a$, and $x = b$, about the *x*-axis as you complete parts (1) through (6).

   (1) Draw a set of axes. Mark the values of the lower and upper bounds of the closed interval $[a,b]$ on the *x*-axis.

   (2) Over $[a,b]$, sketch the graph of *f* (make *f* slightly curved).

   (3) As usual, let $\Delta x = \dfrac{b - a}{n}$. Label $x_0, x_1, x_2, \ldots, x_{i-1}, x_i, \ldots, x_{n-1}$, $x_n$ where $x_0 = a$ and $x_n = b$, and the distance between each successive pair of *x*'s is $\Delta x$.

   (4) Sketch a typical rectangle (use the right endpoint of the subinterval to determine the height of the rectangle).

   (5) As a rectangle revolves around the *x*-axis, it generates a disc. Sketch the disc that a typical rectangle generates.

**Modeling Template**

| I | | | | | | | | | |
|---|---|---|---|---|---|---|---|---|---|
| | | | | | | | | | |
| | | | | | | | | | |
| | | | | | | | | | |
| | | | | | | | | | |
| | | | | | | | | | |

| II | | | | | | | | | |
|---|---|---|---|---|---|---|---|---|---|
| | | | | | | | | | |
| | | | | | | | | | |
| | | | | | | | | | |
| | | | | | | | | | |
| | | | | | | | | | |

**III**

Name _____

**Directions.**

1. Graph $y = f(x)$ in region I.

$y = f(x)$

2. Make an identical graph of $y = f(x)$ in region II.

3. Fold the piece as follows:

4. For the given problem, cut slits up the curve $y = f(x)$ in region II (also cutting into region III at the same time) at the given $x$'s.

5. From a piece of blank cardboard, cut out circles so that the radius of each circle corresponds to the height $y = f(x)$ at the given $x$'s. Use a compass.

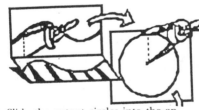

6. Slide the cutout circles into the appropriate slits.

**(6)** Sketch the solid of revolution generated by revolving the region bounded by the $x$-axis, the graph of $f$, $x = a$, and $x = b$, about the $x$-axis.

**b.** Let $D_i$, for $1 \leq i \leq n$, be the disc formed by revolving the $i$th rectangle around the $x$-axis. Find formulas for the volume of the first, second, $i$th, and $n$th discs.

    **(1)** Volume$(D_1)$ =

    **(2)** Volume$(D_2)$ =

    **(3)** Volume$(D_i)$ =

    **(4)** Volume$(D_n)$ =

**c.** Approximate the volume of the solid by finding the sum of the volumes of the discs. Express the result using sigma notation.

**d.** Explain how can you improve the approximation.

**e.** Express the exact volume of the solid of revolution using the appropriate limit.

**f.** Express the exact volume of the solid of revolution in terms of a definite integral.

5. Hopefully, in the last part, you discovered that

$$\text{Volume} = \lim_{\Delta x \to 0} \sum_{i=1}^{n} \pi (f(x_i))^2 \, \Delta x = \int_a^b \pi (f(x))^2 \, dx.$$

Use this formula to find the exact volume of the solids of revolution which you considered in part 2 and again in part 3.

i. Use your CAS to graph the region that generates the solid. Place a copy of the graph in your activity guide. On the graph, sketch a typical rectangle. Sketch the disc generated by the rectangle. Sketch the solid of revolution generated by the region.

ii. Find the volume of the solid by solving the associated integral.

iii. Find the difference between the exact volume and the approximate volume which you calculated in part 3.

a. $f(x) = \sqrt{x}$ for $1 \leq x \leq 9$

b. $f(x) = \sin(x)$ for $0 \leq x \leq \pi$

Note: $\int \sin^2(x) \, dx = \frac{1}{2}x - \frac{1}{4}\sin(2x) + C.$

**c.** $f(x) = \dfrac{1}{x}$ for $1 \le x \le 8$

**d.** $f(x) = x + 1$ for $2 \le x \le 9$

**6.** Does the formula for finding the volume of a solid of revolution using right circular cylinders hold if the graph of $f$ crosses the $x$-axis? For example, does it hold in the case of the solid generated by the region bounded by $f(x) = \sin(x)$ for $-\pi \le x \le \pi$ and the $x$-axis? Justify your response.

In the last task, you considered solids of revolution generated by revolving a region bounded by the $x$-axis, the graph of $f$, $x = a$, and $x = b$, about the $x$-axis. You approximated the volumes of the solids by slicing them into thin right circular cylinders and adding up the volumes of the discs. The more discs you consider, the better the approximation. If you take the limit of the sum of the volumes of the discs as the thickness of the discs gets smaller and smaller, you get the exact volume of the solid. This result can be expressed in terms of a definite integral. In particular, the volume of the solid generated by revolving the region bounded by the $x$-axis, the graph of $f$, $x = a$, $x = b$, about the $x$-axis is given by

$$\text{Volume} = \lim_{\Delta x \to 0} \underbrace{\underbrace{\sum_{i=1}^{n} \underbrace{\pi(f(x_i))^2 \Delta x}_{\text{volume of }i\text{th disc}}}_{\text{approximate volume of solid}} = \int_a^b \pi(f(x))^2 \, dx,}_{\text{volume of solid of revolution}}$$

where $\Delta x = \dfrac{b - a}{n}$ and $x_i = a + i\Delta x$ for $1 \le i \le n$.

Next, consider a solid of revolution generated by revolving a region between two positive curves—an upper curve and a lower curve as in Figure 1. The resulting solid, unlike the solids that you considered before, has a "hole" in it (see Fig. 2).

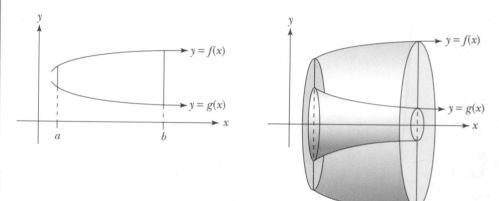

**Figure 1.**                    **Figure 2.**

One way to find the volume of a solid of revolution with a hole is to subtract the volume of the hole—that is, the volume of the solid of revolution generated by the inner curve $y = g(x)$—from the volume of the solid of revolution generated by the outer curve $y = f(x)$:

$$\text{Volume} = \int_a^b \pi(f(x))^2 \, dx - \int_a^b \pi(g(x))^2 \, dx$$

or, combining the two integrals and factoring out $\pi$,

$$\text{Volume} = \int_a^b \pi[(f(x))^2 - (g(x))^2] \, dx.$$

What do these formulas represent geometrically? Observe that if you subdivide the closed interval $[a,b]$ in the usual way, then the rectangle, determined by the right endpoint of each subinterval and extending from the graph of $g$ to the graph of $f$, generates a *washer* or a right circular cylinder with a hole (see Figs. 3 and 4).

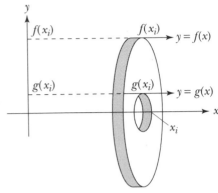

**Figure 3.**                    **Figure 4.**

Since a washer is just a big cylinder whose hole is formed by a smaller cylinder, its volume can be determined by subtracting the volume of the inner cylinder from the volume of the outer cylinder—that is,

$$\text{Volume of washer} = \pi r_{\text{out}}^2 h - \pi r_{\text{in}}^2 h = \pi (r_{\text{out}}^2 - r_{\text{in}}^2) h.$$

Before you calculate any volumes, derive the formula for finding the volume of a solid of revolution with a hole, by summing up the volumes of the associated washers and then passing to the limit; that is, show that

$$\text{Volume} = \int_a^b \pi[f(x))^2 - (g(x))^2] \, dx.$$

---

## Task 9-11: Finding the Volume of a Solid of Revolution Using the Washer Approach

1. Derive the formula for the volume of the solid of revolution by considering the solid generated by the general region $R$ depicted in Figures 3 and 4.

   **a.** Find formulas for the volumes of the first, second, $i$th, and $n$th washers.

   **(1)** Volume$(W_1) =$

   **(2)** Volume$(W_2) =$

   **(3)** Volume$(W_i) =$

   **(4)** Volume$(W_n) =$

**b.** Approximate the volume of the solid of revolution generated by *R* by summing the volumes of the washers. Express your result using sigma notation.

**c.** Find the exact volume by taking the appropriate limit.

**d.** Express the exact volume in terms of a definite integral.

2. In Task 9-10, you drew a region bounded by a single curve and then constructed bases for right circular cylinders which you used to approximate the volume of the solid of revolution generated by the given region. Repeat this process for a solid with a hole.

   Consider the region *R* which is bounded by

   $$f(x) = \sqrt{x} + 2, \qquad g(x) = \frac{1}{4}x + \frac{3}{4}, \qquad x = 1, \quad \text{and} \quad x = 9.$$

   **a.** Construct the solid generated by revolving *R* about the *x*-axis. Use the modeling template, but follow the slightly modified instructions given below.

   *Note: Work together with the members of your group and make one replica of the solid.*

   **Step 1.** Graph the region *R* which generates the solid in region I. Make an identical graph of *R* in region II.

   **Step 2.** Sketch the rectangles which generate the washers used to approximate the volume of the solid determined by *R*. In particular:

   **i.** Let *n* = 8. Subdivide the interval [1,9] into eight equal subintervals.

   **ii.** On *R*, sketch the rectangles determined by the right endpoint of each subinterval and the graphs of the upper and lower curves.

   **Step 3.** For each rectangle, cut out the base of the washer determined by the right endpoint of the associated subinterval and place it on the model. In particular:

   **i.** For each rectangle, cut a washer whose inner radius is determined by the bottom curve and whose outer radius is determined by the top curve.

**ii.** At the right endpoint of each subinterval, cut a slit up to the top curve in region II. Cut into region III at the same time.

**iii.** Place the washer in its appropriate slit.

**b.** Is the volume generated by the washers less than or greater than the actual volume generated by *R*? Why? What happens if you increase the number of washers? Decrease the number?

**c.** Approximate the volume of the solid by summing the volumes of the eight washers.

**d.** Use your CAS to find the actual volume of the solid generated by *R*.

**e.** Do your calculations in parts c and d support your responses to the questions in part b?

In the previous tasks, you thought about how to calculate the volume of a solid of revolution. You estimated the volume by slicing the solid into pieces which were generated by revolving rectangles around the axis. All the slices—or cross sections—had the same shape (either a disc or a washer). However, their sizes varied. Next, think about the case in which the cross sections not only have the same shape but also the same size.

A *cylindrically shaped solid* is a solid such that if you cut it in slices parallel to the base, all the cross sections have the same size and shape as the base itself. For example, if you slice a tin can parallel to its base, all the cross sections are circles which have the same radius as the base of the can. Since the area of any cross section equals the area of the base, the volume of a cylindrically shaped solid can be determined by multiplying the area of its base times its height or thickness; that is,

Volume = (area of the base) · (thickness).

It turns out that some bottles are cylindrically shaped solids. The goal of the next task is to approximate (using several different approaches) how much a container like this holds when it is filled to different levels.

In particular, suppose you would like to know the volume when the level of the liquid is 12, 10, and so on, down to 2 centimeters. After thinking about this for a bit, you say to yourself, "Aha! This bottle is a cylindrically shaped solid! I can find the volume at the various levels by simply finding the area of the base—which, in this case, is the face of the bottle—when it is filled to the specified level and then multiplying the area times the thickness of the bottle." But this approach leads to another puzzle: How can you find the area of the face at the various levels? . . . . Calculus to the rescue! Several possibilities come to mind. You realize you can:

- Use the Trapezoid Rule to approximate the area of the face of the bottle, and multiply the result by the thickness of the bottle.
- Fit a curve to the edge of the bottle and integrate the resulting function over the appropriate bounds to approximate the area of the face of the bottle, and multiply the result by the thickness of the bottle.
- Integrate the curve that generates the face of the bottle and multiply the result by its thickness.
- Fill an empty bottle with water to the given height and weigh it to find its mass and hence its volume.

> Note: mass = density · volume, but since the density of water is 1 gram per cubic centimeter (g/cm³), the numerical value of the mass equals the numerical value of the volume.

Try these approaches in the next task.

---

### Task 9-12: (Project) Filling a Nectar Bottle

Use the four approaches mentioned above to approximate the volume in a bottle that has a circular face and uniform thickness when the height of the liquid is 12, 10, and so on, down to 2 centimeters (cm).

> Note: If you have access to an empty bottle, use it to answer the following questions. If you do not, then use the diagram of the bottle (which appears on the next page) to do parts 1 through 3 and omit part 4.

1. Use the Trapezoid Rule to approximate the area of the face for $h = 2$, 4, 6, 8, 10, and 12 cm, and then use this information to approximate the volume at each level.

   a. Tip the bottle sideways with the cork to your right and trace the upper half of the face of the bottle on a piece of paper so that the $x$-axis corresponds to the center of the bottle and the values on the $x$-axis correspond to the levels.

   b. Divide the interval [0,12] into 12 subintervals of 1 cm each.

   c. Sketch the associated trapezoids.

   d. Use you ruler to determine the $y$-value corresponding to each $x_i$, for $0 \le i \le 13$. (Do your measurements in centimeters.)

   e. Use your CAS to:

   (1) Find the area of each trapezoid.

   (2) Approximate the area of the face for $h = 2$, 4, and so on up to 12, by adding the areas of the appropriate trapezoids.

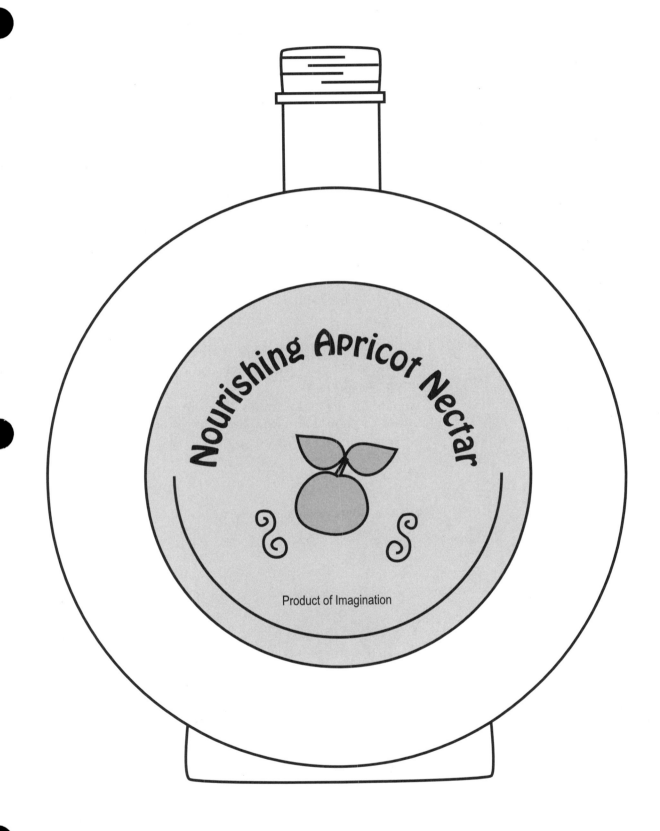

Nourishing Apricot Nectar

Product of Imagination

**(3)** Approximate half the volume of the bottle for $h = 2$, 4, and so on, up to 12 cm, by multiplying the results from part (2) by the thickness of the bottle.

**(4)** Approximate the volume by multiplying the results from part (3) by 2.

**f.** Enter your results in the table below.

| Height (cm) | 2 | 4 | 6 | 8 | 10 | 12 |
|---|---|---|---|---|---|---|
| Volume (cm³) | | | | | | |

**2.** Approximate the volume again, only this time fit a curve to the edge of the bottle. Approximate the areas corresponding to $h = 2$, 4, 6, 8, 10, and 12 cm by integrating this function over the appropriate intervals. Use the resulting values to estimate the volume at the various levels.

**a.** Use your CAS to:

**(1)** Find a polynomial that closely approximates the curve.

**(2)** Approximate the desired areas by integrating the polynomial over the appropriate domains.

**(3)** Approximate the desired volumes by multiplying the areas by the thickness of the bottle and doubling the results.

**b.** Print a copy of your CAS session, and place it below. Label your results.

**3.** Try another approach. Assume the face of the bottle is a circle. Find the areas corresponding to $h = 2$, 4, 6, 8, 10, and 12 cm by integrating the function that generates the face of the bottle over the appropriate intervals. Use the resulting values to find the volume at the various levels.

**a.** Find the radius of the circle and its center. Use this information to find the equation for the curve that generates the face.

Recall that the equation of a circle with center $C(a,b)$ and radius $r$ is given by

$$(x - a)^2 + (y - b)^2 = r^2.$$

To make things a little simpler, you might want to shift the location of the bottle so that the center of its face is at the origin. If you do this, be sure and also shift the values of the various heights.

**b.** Use your CAS to:

**(1)** Graph the upper half of the circle.

**(2)** Find the desired areas by integrating the function over the appropriate domains.

**(3)** Find the desired volumes by multiplying the areas by the thickness of the bottle and doubling the results.

**c.** Print a copy of your CAS session, and place it below. Label your results.

**4.** One more time. If you have access to an empty bottle, find the mass at each level and use this information to calculate the volume.

Recall that the numerical value of the mass equals the numerical value of the volume, since

$$\text{mass} = (\text{density}) \cdot (\text{volume})$$

where the density of water is $1 \text{ g/cm}^3$.

Fill the bottle with water to the designated levels and use a scale to find the mass at each level. Enter your results in the table below.

| Height (cm) | 2 | 4 | 6 | 8 | 10 | 12 |
|---|---|---|---|---|---|---|
| Mass (g) | | | | | | |

**5.** Compare your results from parts 1 to 4. Also compare the results of your group with the other groups in your class. Are there any major discrepancies? If so, try to explain why.

## Unit 9 Homework After Section 3

- Complete the tasks in Section 3. Be prepared to discuss them in class.

- Solve some improper integrals in HW9.11.

**HW9.11**

1. In each of the following improper integrals, at least one of the limits of integration is infinite. Graph the function and shade the region associated with the integral. Determine if the integral converges or if it diverges. If it converges, find its value. If it diverges, show why this is the case.

   **a.** $\int_0^\infty e^{-x}\, dx$

   **b.** $\int_2^\infty \dfrac{1}{x-1}\, dx$

   **c.** $\int_2^\infty \dfrac{1}{(x-1)^2}\, dx$

   **d.** $\int_{-\infty}^1 e^x\, dx$

   **e.** $\int_1^\infty e^x\, dx$

   **f.** $\int_0^\infty \dfrac{x}{1+x^2}\, dx$

   **g.** $\int_{-\infty}^0 \dfrac{1}{(x-3)^3}\, dx$

   **h.** $\int_{-\infty}^\infty xe^x\, dx$

   **i.** $\int_{-\infty}^2 \dfrac{5}{5-2x}\, dx$

   **j.** $\int_{-\infty}^0 \dfrac{1}{(4-x)^4}\, dx$

2. In each of the following improper integrals, the function has a blowup discontinuity at one of the limits of integration. Graph the function and shade the region associated with the integral. Determine if the integral converges or if it diverges. If it converges, find its value. If it diverges, show why this is the case.

   **a.** $\int_{-4}^{-1} \dfrac{1}{(x+1)^3}\, dx$

   **b.** $\int_0^1 \dfrac{e^{\sqrt{x}}}{\sqrt{x}}\, dx$

   **c.** $\int_0^4 \dfrac{1}{x^2}\, dx$

   **d.** $\int_0^8 \dfrac{1}{\sqrt[3]{x}}\, dx$

   **e.** $\int_{-2}^0 \dfrac{x}{\sqrt{4-x^2}}\, dx$

   **f.** $\int_0^3 \dfrac{1}{\sqrt{3-x}}\, dx$

3. Consider $\int_0^4 \dfrac{1}{(x-3)^2}\, dx$.

   **a.** At first glance, it appears that you can use the Fundamental Theorem of Calculus to solve this integral. Try it.

   **b.** Graph the function and shade the region associated with the integral. Use your graph to explain why the "answer" you calculated in part a is not reasonable.

   **c.** Explain why the Fundamental Theorem does not apply in this case. In particular, which condition does not hold?

**d.** In general, if $f$ has a blowup discontinuity at $x = c$, where $a \leq c \leq b$, but $f$ is continuous everywhere else on the interval $[a,b]$, then

$$\int_a^b f(x) \; dx = \int_a^c f(x) \; dx + \int_c^b f(x) \; dx$$

converges, provided both the integrals on the right side of the equation converge.

**(1)** Express $\displaystyle\int_0^4 \frac{1}{(x-3)^2} \; dx$ as the sum of two integrals, with $c = 3$.

**(2)** Determine whether $\displaystyle\int_0^4 \frac{1}{(x-3)^2} \; dx$ converges or diverges.

4. Compare two regions that have similar shapes.

   **a.** Find the area under the graph of $y = (t-1)^{-3/2}$, where $t \geq 3$.

   **b.** Show that the area of the region under the graph of $y = (t-1)^{-1/3}$, where $t \geq 3$, cannot be assigned a finite number.

5. Consider some general situations.

   **a.** Show that $\displaystyle\int_1^\infty \frac{1}{x^n} \; dx$ converges if $n > 1$ and diverges if $n \leq 1$.

   **b.** Show that $\displaystyle\int_0^1 \frac{1}{x^n} \; dx$ converges if $n < 1$ and diverges if $n \geq 1$.

- Find the volumes of some solids of revolution using the disc approach in HW9.12 and HW9.13.

**HW9.12** In each of the regions in parts 1 through 3, use your CAS to:

   **i.** Plot the region bounded by the graph of the given equation and the $x$-axis.

   **ii.** Find the volume of the solid generated by revolving the region about the $x$-axis.

Print a copy of your CAS session. On each graph, sketch a typical rectangle and the disc that it generates.

1. $y = x^3 + 1$ for $0 \leq x \leq 2$

2. $g(x) = e^x$ for $-1 \leq x \leq 1$

3. $f(x) = \cos(x)$ for $-\frac{\pi}{2} \leq x \leq \frac{3\pi}{2}$

**HW9.13** Find the volumes of the solids described in Task 9-10, part 1. Do your calculations by hand.

1. The region is the first quadrant bounded by $x = 2$, $x = 5.5$, and $y = 3.25$. The $x$-axis is the axis of revolution.

2. The region bounded by the semicircle located in the first and second quadrants that has radius 3 and center at the origin. The $x$-axis is the axis of revolution.

3. The region bounded by $y = x^2 + 1$, where $-2 \le x \le 2$, and the $x$-axis. The $x$-axis is the axis of revolution.

4. The region bounded by $f(x) = \sin(x)$ for $-\pi \le x \le \pi$ and the $x$-axis. The $x$-axis is the axis of revolution.

• Use the disc approach to prove some familiar formulas in HW9.14.

**HW9.14** Use the disc approach to prove the formulas in parts 1 through 3 for finding the volume of some regular solids. In each case:

    **i.** Sketch a region $R$ which generates the solid when revolved about the $x$-axis.

    **ii.** Label the graphs which bound $R$.

    **iii.** Solve the definite integral that gives the volume of the solid.

1. The volume of a right circular cylinder of altitude $h$ and radius of base $r$ is $\pi r^2 h$.

2. The volume of a right circular cone of altitude $h$ and radius of base $r$ is $\frac{1}{3}\pi r^2 h$.

3. The volume of a sphere of radius $r$ is $\frac{4}{3}\pi r^3$.

• Use the washer approach to find the volumes of some solids of revolution in HW9.15.

**HW9.15**

1. For each of the regions in parts a and b:

    **i.** Use your CAS to:

    • Plot the region $R$ bounded by the graphs of the given curves.

    • Find the volume of the solid generated by revolving $R$ about the $x$-axis.

    **ii.** Print a copy of your CAS session. On each graph, sketch a typical rectangle and the washer it generates.

    **a.** $y = \sin(x) + 2$, $y = 2$, $x = 0$, and $x = \pi$

    **b.** $y = x^2 + 2$ and $y = 3x + 2$

2. For each of the regions in parts a through c, use pencil and paper to do the following:

    **i.** Sketch the region $R$ bounded by the graphs of the given curves.

    **ii.** Sketch a typical rectangle and the washer generated by the rectangle.

    **iii.** Find the volume of the solid generated by revolving $R$ about the $x$-axis.

    **a.** $y = e^x$, $y = 1$, $x = 0$, and $x = 2$

    **b.** $y = x^2 + 6$ and $y = -x^2 + 4$, where $-2 \le x \le 2$

• Write your journal entry for this unit. As usual, before you begin to write, review the material in the unit. Think about how it all fits together. Try to identify what, if anything, is still causing you trouble.

**HW9.16** Write your journal entry for Unit 9.

1. Reflect on what you have learned in this unit. Describe in your own words the concepts that you studied and what you learned about them. How do they fit together? What concepts were easy? Hard? What were the main (important) ideas? Give some examples of the main ideas.

2. Reflect on the learning environment for the course. Describe the aspects of this unit and the learning environment that helped you understand the concepts you studied. What activities did you like? Dislike?

# *Appendix:*
## TABLE OF INTEGRALS

**148.** In the following table, the constant of integration, $C$, is omitted but should be added to the result of every integration. The letter $x$ represents any variable; $u$ represents any function of $x$; the remaining letters represent arbitrary constants, unless otherwise indicated; all angles are in radians. Unless otherwise mentioned, $\log_e u = \log u$.

## Expressions Containing $ax + b$

**1.** $\displaystyle\int (ax + b)^n \, dx = \frac{1}{a(n+1)}(ax + b)^{n+1}, \quad n \neq -1.$

**2.** $\displaystyle\int \frac{dx}{ax + b} = \frac{1}{a}\log_e(ax + b).$

**3.** $\displaystyle\int \frac{dx}{(ax + b)^2} = -\frac{1}{a(ax + b)}.$

**4.** $\displaystyle\int \frac{dx}{(ax + b)^3} = -\frac{1}{2a(ax + b)^2}.$

**5.** $\displaystyle\int x(ax + b)^n \, dx = \frac{1}{a^2(n+2)}(ax + b)^{n+2}$
$$-\frac{b}{a^2(n+1)}(ax + b)^{n+1}, \quad n \neq -1, -2.$$

**6.** $\displaystyle\int \frac{x \, dx}{ax + b} = \frac{x}{a} - \frac{b}{a^2}\log(ax + b).$

**7.** $\displaystyle\int \frac{x \, dx}{(ax + b)^2} = \frac{b}{a^2(ax + b)} + \frac{1}{a^2}\log(ax + b).$

8. $\int \dfrac{x\,dx}{(ax+b)^3} = \dfrac{b}{2a^2(ax+b)^2} - \dfrac{1}{a^2(ax+b)}.$

9. $\int x^2(ax+b)^n\,dx = \dfrac{1}{a^2}\left(\dfrac{(ax+b)^{n+3}}{n+3} - 2b\dfrac{(ax+b)^{n+2}}{n+2}\right.$

$\left. + \dfrac{b^2(ax+b)^{n+1}}{n+1}\right), \quad n \neq -1, -2, -3.$

10. $\int \dfrac{x^2\,dx}{ax+b} = \dfrac{1}{a^3}\left(\dfrac{1}{2}(ax+b)^2 - 2b(ax+b) + b^2\log(ax+b)\right).$

11. $\int \dfrac{x^2\,dx}{(ax+b)^2} = \dfrac{1}{a^3}\left((ax+b) - 2b\log(ax+b) - \dfrac{b^2}{ax+b}\right).$

12. $\int \dfrac{x^2\,dx}{(ax+b)^3} = \dfrac{1}{a^3}\left(\log(ax+b) + \dfrac{2b}{ax+b} - \dfrac{b^2}{2(ax+b)^2}\right).$

13. $\int x^m(ax+b)^n\,dx$

$= \dfrac{1}{a(m+n+1)}\left(x^m(ax+b)^{n+1} - mb\!\int x^{m-1}(ax+b)^n\,dx\right)$

$= \dfrac{1}{m+n+1}\left(x^{m+1}(ax+b)^n + nb\!\int x^m(ax+b)^{n-1}\,dx\right),$

$m > 0,\ m+n+1 \neq 0.$

14. $\int \dfrac{dx}{x(ax+b)} = \dfrac{1}{b}\log\left(\dfrac{x}{ax+b}\right).$

15. $\int \dfrac{dx}{x^2(ax+b)} = -\dfrac{1}{bx} + \dfrac{a}{b^2}\log\left(\dfrac{ax+b}{x}\right).$

16. $\int \dfrac{dx}{x^3(ax+b)} = \dfrac{2ax-b}{2b^2x^2} + \dfrac{a^2}{b^3}\log\left(\dfrac{x}{ax+b}\right).$

17. $\int \dfrac{dx}{x(ax+b)^2} = \dfrac{1}{b(ax+b)} - \dfrac{1}{b^2}\log\left(\dfrac{ax+b}{x}\right).$

18. $\int \dfrac{dx}{x(ax+b)^3} = \dfrac{1}{b^3}\left[\dfrac{1}{2}\left(\dfrac{ax+2b}{ax+b}\right)^2 + \log\left(\dfrac{x}{ax+b}\right)\right].$

19. $\int \dfrac{dx}{x^2(ax+b)^2} = -\dfrac{b+2ax}{b^2x(ax+b)} + \dfrac{2a}{b^3}\log\left(\dfrac{ax+b}{x}\right).$

20. $\int \sqrt{ax+b}\,dx = \dfrac{2}{3a}\sqrt{(ax+b)^3}.$

21. $\int x\sqrt{ax+b}\,dx = \dfrac{2(3ax-2b)}{15a^2}\sqrt{(ax+b)^3}.$

22. $\int x^2\sqrt{ax+b}\,dx = \dfrac{2(15a^2x^2 - 12abx + 8b^2)\sqrt{(ax+b)^3}}{105a^3}$

23. $\int x^3\sqrt{ax+b}\,dx = \dfrac{2(35a^3x^3 - 30a^2bx^2 + 24ab^2x - 16b^3)\sqrt{(ax+b)^3}}{315a^4}.$

**24.** $\int x^n \sqrt{ax + b}\, dx = \dfrac{2}{a^{n+1}} \int u^2(u^2 - b)^n\, du, \qquad u = \sqrt{ax + b}.$

**25.** $\int \dfrac{\sqrt{ax + b}}{x}\, dx = 2\sqrt{ax + b} + b \int \dfrac{dx}{x\sqrt{ax + b}}.$

**26.** $\int \dfrac{dx}{\sqrt{ax + b}} = \dfrac{2\sqrt{ax + b}}{a}.$

**27.** $\int \dfrac{x\, dx}{\sqrt{ax + b}} = \dfrac{2(ax - 2b)}{3a^2} \sqrt{ax + b}.$

**28.** $\int \dfrac{x^2\, dx}{\sqrt{ax + b}} = \dfrac{2(3a^2x^2 - 4abx + 8b^2)}{15a^3} \sqrt{ax + b}.$

**29.** $\int \dfrac{x^3\, dx}{\sqrt{ax + b}} = \dfrac{2(5a^3x^3 - 6a^2bx^2 + 8ab^2x - 16b^3)}{35a^4} \sqrt{ax + b}.$

**30.** $\int \dfrac{x^n\, dx}{\sqrt{ax + b}} = \dfrac{2}{a^{n+1}} \int (u^2 - b)^n\, du, \qquad u = \sqrt{ax + b}.$

**31.** $\int \dfrac{dx}{x\sqrt{ax + b}} = \dfrac{1}{\sqrt{b}} \log\left(\dfrac{\sqrt{ax + b} - \sqrt{b}}{\sqrt{ax + b} + \sqrt{b}}\right) \quad \text{for } b > 0.$

**32.** $\int \dfrac{dx}{x\sqrt{ax + b}} = \dfrac{2}{\sqrt{-b}} \tan^{-1} \sqrt{\dfrac{ax + b}{-b}}, \quad b < 0$

$\qquad\qquad\quad = \dfrac{-2}{\sqrt{b}} \tanh^{-1} \sqrt{\dfrac{ax + b}{-b}}, \quad b > 0.$

**33.** $\int \dfrac{dx}{x^2\sqrt{ax + b}} = -\dfrac{\sqrt{ax + b}}{bx} - \dfrac{a}{2b} \int \dfrac{dx}{x\sqrt{ax + b}}.$

**34.** $\int \dfrac{dx}{x^3\sqrt{ax + b}} = -\dfrac{\sqrt{ax + b}}{2bx^2} + \dfrac{3a\sqrt{ax + b}}{4b^2x} + \dfrac{3a^2}{8b^2} \int \dfrac{dx}{x\sqrt{ax + b}}.$

**35.** $\int \dfrac{dx}{x^n(ax + b)^m} = -\dfrac{1}{b^{m+n-1}} \int \dfrac{(u - a)^{m+n-2}\, du}{u^m}, \qquad u = \dfrac{ax + b}{x}.$

**36.** $\int (ax + b)^{\pm n/2}\, dx = \dfrac{2(ax + b)^{(2\pm n)/2}}{a(2 \pm n)}.$

**37.** $\int x(ax + b)^{\pm n/2}\, dx = \dfrac{2}{a^2}\left(\dfrac{(ax + b)^{(4\pm n)/2}}{4 \pm n} - \dfrac{b(ax + b)^{(2\pm n)/2}}{2 \pm n}\right).$

**38.** $\int \dfrac{dx}{x(ax + b)^{n/2}} = \dfrac{1}{b} \int \dfrac{dx}{x(ax + b)^{(n-2)/2}} - \dfrac{a}{b} \int \dfrac{dx}{(ax + b)^{n/2}}.$

**39.** $\int \dfrac{x^m\, dx}{\sqrt{ax + b}} = \dfrac{2x^m\sqrt{ax + b}}{(2m + 1)a} - \dfrac{2mb}{(2m + 1)a} \int \dfrac{x^{m-1}\, dx}{\sqrt{ax + b}}.$

**40.** $\int \dfrac{dx}{x^n\sqrt{ax + b}} = \dfrac{-\sqrt{ax + b}}{(n - 1)bx^{n-1}} - \dfrac{(2n - 3)a}{(2n - 2)b} \int \dfrac{dx}{x^{n-1}\sqrt{ax + b}}.$

**41.** $\int \dfrac{(ax + b)^{n/2}}{x}\, dx = a \int (ax + b)^{(n-2)/2}\, dx + b \int \dfrac{(ax + b)^{(n-2)/2}}{x}\, dx.$

**42.** $\int \dfrac{dx}{(ax+b)(cx+d)} = \dfrac{1}{bc-ad} \log\left(\dfrac{cx+d}{ax+b}\right), \qquad bc - ad \neq 0.$

**43.** $\int \dfrac{dx}{(ax+b)^2(cx+d)}$

$$= \dfrac{1}{bc-ad}\left[\dfrac{1}{ax+b} + \dfrac{c}{bc-ad}\log\left(\dfrac{cx+d}{ax+b}\right)\right], \qquad bc - ad \neq 0.$$

**44.** $\int (ax+b)^n(cx+d)^m\,dx = \dfrac{1}{(m+n+1)a}\left[(ax+b)^{n+1}(cx+d)^m\right.$

$$\left. - m(bc-ad)\int (ax+b)^n(cx+d)^{m-1}\,dx\right].$$

**45.** $\int \dfrac{dx}{(ax+b)^n(cx+d)^m} = \dfrac{1}{(m-1)(bc-ad)}\left(\dfrac{1}{(ax+b)^{n-1}(cx+d)^{m-1}}\right.$

$$\left. - a(m+n-2)\int \dfrac{dx}{(ax+b)^n(cx+d)^{m-1}}\right), \qquad m > 1,\ n > 0,\ bc - ad \neq 0.$$

**46.** $\int \dfrac{(ax+b)^n}{(cx+d)^m}\,dx$

$$= -\dfrac{1}{(m-1)(bc-ad)}\left(\dfrac{(ax+b)^{n+1}}{(cx+d)^{m-1}} + (m-n-2)a\int \dfrac{(ax+b)^n\,dx}{(cx+d)^{m-1}}\right)$$

$$= \dfrac{1}{(m-n-1)c}\left(\dfrac{(ax+b)^n}{(cx+d)^{m-1}} + n(bc-ad)\int \dfrac{(ax+b)^{n-1}}{(cx+d)^m}\,dx\right).$$

**47.** $\int \dfrac{x\,dx}{(ax+b)(cx+d)} = \dfrac{1}{bc-ad}\left(\dfrac{b}{a}\ln|ax+b|\right.$

$$\left. - \dfrac{d}{c}\log(cx+d)\right), \qquad bc - ad \neq 0.$$

**48.** $\int \dfrac{x\,dx}{(ax+b)^2(cx+d)} = \dfrac{1}{bc-ad}\left(-\dfrac{b}{a(ax+b)}\right.$

$$\left. - \dfrac{d}{bc-ad}\log\left(\dfrac{cx+d}{ax+b}\right)\right), \qquad bc - ad \neq 0.$$

**49.** $\int \dfrac{cx+d}{\sqrt{ax+b}}\,dx = \dfrac{2}{3a^2}(3ad - 2bc + acx)\sqrt{ax+b}.$

**50.** $\int \dfrac{\sqrt{ax+b}}{cx+d}\,dx = \dfrac{2\sqrt{ax+b}}{c}$

$$- \dfrac{2}{c}\sqrt{\dfrac{ad-bc}{c}}\tan^{-1}\sqrt{\dfrac{c(ax+b)}{ad-bc}}, \qquad c > 0,\ ad > bc.$$

**51.** $\int \dfrac{\sqrt{ax+b}}{cx+d}\,dx = \dfrac{2\sqrt{ax+b}}{c}$

$$+ \dfrac{1}{c}\sqrt{\dfrac{bc-ad}{c}}\log\left(\dfrac{\sqrt{c(ax+b)} - \sqrt{bc-ad}}{}\right), \qquad c > 0,\ bc > ad.$$

**52.** $\int \dfrac{dx}{(cx+d)\sqrt{ax+b}} = \dfrac{2}{\sqrt{c}\sqrt{ad-bc}}\tan^{-1}\sqrt{\dfrac{c(ax+b)}{ad-bc}}, \qquad c > 0,\ ad > bc.$

**53.** $\displaystyle\int \frac{dx}{(cx + d)\sqrt{ax + b}}$

$$= \frac{1}{\sqrt{c}\sqrt{bc - ad}} \log\left(\frac{\sqrt{c(ax + b)} - \sqrt{bc - ad}}{}\right), \qquad c > 0, \ bc > ad.$$

## Expressions Containing $ax^2 + c$, $ax^n + c$, $x^2 \pm p^2$, and $p^2 - x^2$

**54.** $\displaystyle\int \frac{dx}{p^2 + x^2} = \frac{1}{p} \tan^{-1}\frac{x}{p} \quad \text{or} \quad -\frac{1}{p} \operatorname{ctn}^{-1}\left(\frac{x}{p}\right).$

**55.** $\displaystyle\int \frac{dx}{p^2 - x^2} = \frac{1}{2p} \log\left(\frac{p + x}{p - x}\right), \quad \text{or} \quad \frac{1}{p} \tanh^{-1}\left(\frac{x}{p}\right).$

**56.** $\displaystyle\int \frac{dx}{ax^2 + c} = \frac{1}{\sqrt{ac}} \tan^{-1}\left(x\sqrt{\frac{a}{c}}\right), \qquad a \text{ and } c > 0.$

**57.** $\displaystyle\int \frac{dx}{ax^2 + c} = \frac{1}{2\sqrt{-ac}} \log\left(\frac{x\sqrt{a} - \sqrt{-c}}{}\right), \qquad n > 0, \ c < 0$

$$= \frac{1}{2\sqrt{-ac}} \log\left(\frac{\sqrt{c} + x\sqrt{-a}}{\sqrt{c} - x\sqrt{-a}}\right), \qquad a < 0, \ c > 0.$$

**58.** $\displaystyle\int \frac{dx}{(ax^2 + c)^n} = \frac{1}{2(n - 1)c} \frac{x}{(ax^2 + c)^{n-1}}$

$$+ \frac{2n - 3}{2(n - 1)c} \int \frac{dx}{(ax^2 + c)^{n-1}}, \qquad n \text{ a positive integer.}$$

**59.** $\displaystyle\int x(ax^2 + c)^n \, dx = \frac{1}{2a} \frac{(ax^2 + c)^{n+1}}{n + 1}, \qquad n \neq -1.$

**60.** $\displaystyle\int \frac{x}{ax^2 + c} \, dx = \frac{1}{2a} \log(ax^2 + c).$

**61.** $\displaystyle\int \frac{dx}{x(ax^2 + c)} = \frac{1}{2c} \log\left(\frac{ax^2}{ax^2 + c}\right).$

**62.** $\displaystyle\int \frac{dx}{x^2(ax^2 + c)} = -\frac{1}{cx} - \frac{a}{c} \int \frac{dx}{ax^2 + c}.$

**63.** $\displaystyle\int \frac{x^2 \, dx}{ax^2 + c} = \frac{x}{a} - \frac{c}{a} \int \frac{dx}{ax^2 + c}.$

**64.** $\displaystyle\int \frac{x^n \, dx}{ax^2 + c} = \frac{x^{n-1}}{a(n - 1)} - \frac{c}{a} \int \frac{x^{n-2} \, dx}{ax^2 + c}, \qquad n \neq 1.$

**65.** $\displaystyle\int \frac{x^2 \, dx}{(ax^2 + c)^n} = -\frac{1}{2(n - 1)a} \frac{x}{(ax^2 + c)^{n-1}}$

$$+ \frac{1}{2(n - 1)a} \int \qquad\qquad \cdot \qquad \frac{dx}{(ax^2 + c)^{n-1}}$$

**66.** $\displaystyle\int\frac{dx}{x^2(ax^2+c)^n} = \frac{1}{c}\int\frac{dx}{x^2(ax^2+c)^{n-1}} - \frac{a}{c}\int\frac{dx}{(ax^2+c)^n}.$

**67.** $\displaystyle\int\sqrt{x^2\pm p^2}\,dx = \frac{1}{2}[x\sqrt{x^2\pm p^2}\pm p^2\log(x+\sqrt{x^2\pm p^2})].$

**68.** $\displaystyle\int\sqrt{p^2-x^2}\,dx = \frac{1}{2}\left[x\sqrt{p^2-x^2}+p^2\sin^{-1}\left(\frac{x}{p}\right)\right].$

**69.** $\displaystyle\int\frac{dx}{\sqrt{x^2\pm p^2}} = \log(x+\sqrt{x^2\pm p^2}).$

**70.** $\displaystyle\int\frac{dx}{\sqrt{p^2-x^2}} = \sin^{-1}\left(\frac{x}{p}\right) \quad\text{or}\quad -\cos^{-1}\left(\frac{x}{p}\right).$

**71.** $\displaystyle\int\sqrt{ax^2+c}\,dx = \frac{x}{2}\sqrt{ax^2+c}$
$$+ \frac{c}{2\sqrt{a}}\log(x\sqrt{a}+\sqrt{ax^2+c}), \qquad a>0.$$

**72.** $\displaystyle\int\sqrt{ax^2+c}\,dx = \frac{x}{2}\sqrt{ax^2+c}+\frac{c}{2\sqrt{-a}}\sin^{-1}\left(x\sqrt{\frac{-a}{c}}\right), \qquad a<0.$

**73.** $\displaystyle\int\frac{dx}{\sqrt{ax^2+c}} = \frac{1}{\sqrt{a}}\log(x\sqrt{a}+\sqrt{ax^2+c}), \qquad a>0.$

**74.** $\displaystyle\int\frac{dx}{\sqrt{ax^2+c}} = \frac{1}{\sqrt{-a}}\sin^{-1}\left(x\sqrt{\frac{-a}{c}}\right), \qquad a<0.$

**75.** $\displaystyle\int x\sqrt{ax^2+c}\,dx = \frac{1}{3a}(ax^2+c)^{3/2}.$

**76.** $\displaystyle\int x^2\sqrt{ax^2+c}\,dx = \frac{x}{4a}\sqrt{(ax^2+c)^3}-\frac{cx}{8a}\sqrt{ax^2+c}$
$$- \frac{c^2}{8\sqrt{a^3}}\log(x\sqrt{a}+\sqrt{ax^2+c}), \qquad a>0.$$

**77.** $\displaystyle\int x^2\sqrt{ax^2+c}\,dx = \frac{x}{4a}\sqrt{(ax^2+c)^3}-\frac{cx}{8a}\sqrt{ax^2+c}$
$$- \frac{c^2}{8a\sqrt{-a}}\sin^{-1}\left(x\sqrt{\frac{-a}{c}}\right), \qquad a<0.$$

**78.** $\displaystyle\int\frac{x\,dx}{\sqrt{ax^2+c}} = \frac{1}{a}\sqrt{ax^2+c}.$

**79.** $\displaystyle\int\frac{x^2\,dx}{\sqrt{ax^2+c}} = \frac{x}{a}\sqrt{ax^2+c}-\frac{1}{a}\int\sqrt{ax^2+c}\,dx.$

**80.** $\displaystyle\int\frac{\sqrt{ax^2+c}}{x}\,dx = \sqrt{ax^2+c}+\sqrt{c}\log\left(\frac{\sqrt{ax^2+c}-\sqrt{c}}{x}\right), \qquad c>0.$

**81.** $\displaystyle\int\frac{\sqrt{ax^2+c}}{x}\,dx = \sqrt{ax^2+c}-\sqrt{-c}\tan^{-1}\frac{\sqrt{ax^2+c}}{\sqrt{-c}}, \qquad c<0.$

**82.** $\displaystyle\int\frac{dx}{x\sqrt{p^2\pm x^2}} = -\frac{1}{p}\log\left(\frac{p+\sqrt{p^2\pm x^2}}{x}\right).$

**83.** $\displaystyle\int \frac{dx}{x\sqrt{x^2 - p^2}} = \frac{1}{p}\cos^{-1}\left(\frac{p}{x}\right)$ or $-\frac{1}{p}\sin^{-1}\left(\frac{p}{x}\right)$.

**84.** $\displaystyle\int \frac{dx}{x\sqrt{ax^2 + c}} = \frac{1}{\sqrt{c}}\log\left(\frac{\sqrt{ax^2 + c} - \sqrt{c}}{x}\right), \qquad c > 0.$

**85.** $\displaystyle\int \frac{dx}{x\sqrt{ax^2 + c}} = \frac{1}{\sqrt{-c}}\sec^{-1}\left(x\sqrt{-\frac{a}{c}}\right), \qquad c < 0.$

**86.** $\displaystyle\int \frac{dx}{x^2\sqrt{ax^2 + c}} = -\frac{\sqrt{ax^2 + c}}{cx}.$

**87.** $\displaystyle\int \frac{x^n\, dx}{\sqrt{ax^2 + c}} = \frac{x^{n-1}\sqrt{ax^2 + c}}{na} - \frac{(n-1)c}{na}\int \frac{x^{n-2}\, dx}{\sqrt{ax^2 + c}}, \qquad n > 0.$

**88.** $\displaystyle\int x^n\sqrt{ax^2 + c}\, dx = \frac{x^{n-1}(ax^2 + c)^{3/2}}{(n+2)a}$

$$- \frac{(n-1)c}{(n+2)a}\int x^{n-2}\sqrt{ax^2 + c}\, dx, \qquad n > 0.$$

**89.** $\displaystyle\int \frac{\sqrt{ax^2 + c}}{x^n}\, dx = -\frac{(ax^2 + c)^{3/2}}{c(n-1)x^{n-1}}$

$$- \frac{(n-4)a}{(n-1)c}\int \frac{\sqrt{ax^2 + c}}{x^{n-2}}\, dx, \qquad n > 1.$$

**90.** $\displaystyle\int \frac{dx}{x^n\sqrt{ax^2 + c}} = -\frac{\sqrt{ax^2 + c}}{c(n-1)x^{n-1}}$

$$- \frac{(n-2)\, a}{(n-1)\, c}\int \frac{dx}{x^{n-2}\sqrt{ax^2 + c}}, \qquad n > 1.$$

**91.** $\displaystyle\int (ax^2 + c)^{3/2}\, dx = \frac{x}{8}(2ax^2 + 5c)\sqrt{ax^2 + c}$

$$+ \frac{3c^2}{8\sqrt{a}}\log(x\sqrt{a} + \sqrt{ax^2 + c}), \qquad a > 0.$$

**92.** $\displaystyle\int (ax^2 + c)^{3/2}\, dx = \frac{x}{8}(2ax^2 + 5c)\sqrt{ax^2 + c}$

$$+ \frac{3c^2}{8\sqrt{-a}}\sin^{-1}\left(x\sqrt{\frac{-a}{c}}\right), \qquad a < 0.$$

**93.** $\displaystyle\int \frac{dx}{(ax^2 + c)^{3/2}} = \frac{x}{c\sqrt{ax^2 + c}}.$

**94.** $\displaystyle\int x(ax^2 + c)^{3/2}\, dx = \frac{1}{5a}(ax^2 + c)^{3/2}.$

**95.** $\displaystyle\int x^2(ax^2 + c)^{3/2}\, dx = \frac{x^3}{6}(ax^2 + c)^{3/2} + \frac{c}{2}\int x^2\sqrt{ax^2 + c}\, dx.$

**96.** $\displaystyle\int x^n(ax^2 + c)^{3/2}\, dx = \frac{x^{n+1}(ax^2 + c)^{3/2}}{n + 4} + \frac{3c}{n + 4}\int x^n\sqrt{ax^2 + c}\, dx.$

## Expressions Containing tan ax or ctn ax
## (tan ax = 1/ctn ax)

**206.** $\int \tan u \, du = -\log(\cos u)$   or   $\log(\sec u)$,   where $u$ is any function of $x$.

**207.** $\int \tan ax \, dx = -\dfrac{1}{a} \log(\cos ax)$.

**208.** $\int \tan^2 ax \, dx = \dfrac{1}{a} \tan ax - x$.

**209.** $\int \tan^3 ax \, dx = \dfrac{1}{2a} \tan^2 ax + \dfrac{1}{a} \log(\cos ax)$.

**210.** $\int \tan^n ax \, dx = \dfrac{1}{a(n-1)} \tan^{n-1} ax - \int \tan^{n-2} ax \, dx$,      $n$ integer $> 1$.

**211.** $\int \text{ctn } u \, du = \log(\sin u)$   or   $-\log(\csc u)$,   where $u$ is any function of $x$.

**212.** $\int \text{ctn}^2 ax \, dx = \int \dfrac{dx}{\tan^2 ax} = -\dfrac{1}{a} \text{ctn } ax - x$.

**213.** $\int \text{ctn}^3 ax \, dx = -\dfrac{1}{2a} \text{ctn}^2 ax - \dfrac{1}{a} \log(\sin ax)$.

**214.** $\int \text{ctn}^n ax \, dx = \int \dfrac{dx}{\tan^n ax}$

$$-\dfrac{1}{a(n-1)} \text{ctn}^{n-1} ax - \int \text{ctn}^{n-2} ax \, dx, \; n \text{ integer} > 1.$$

**215.** $\int \dfrac{dx}{b + c \tan ax} = \int \dfrac{\text{ctn } ax \, dx}{b \, \text{ctn } ax + c}$

$$= \dfrac{1}{b^2 + c^2} \left[ bx + \dfrac{c}{a} \log(b \cos ax + c \sin ax) \right].$$

**216.** $\int \dfrac{dx}{b + c \, \text{ctn } ax} = \int \dfrac{\tan ax \, dx}{b \tan ax + c}$

$$= \dfrac{1}{b^2 + c^2} \left( bx - \dfrac{c}{a} \log(c \cos ax + b \sin ax) \right).$$

**217.** $\int \dfrac{dx}{\sqrt{b + c \tan^2 ax}} = \dfrac{1}{a\sqrt{b-c}} \sin^{-1}\left( \sqrt{\dfrac{b-c}{b}} \sin ax \right)$,      $b$ pos., $b^2 > c^2$.

## Expressions Containing sec *ax* = 1/cos *ax* or csc *ax* = 1/sin *ax*

**218.** $\int \sec u \, du = \log(\sec u + \tan u) = \log\left[\tan\left(\dfrac{u}{2} + \dfrac{\pi}{4}\right)\right],$

where $u$ is any function of $x$.

**219.** $\int \sec ax \, dx = \dfrac{1}{a} \log\left[\tan\left(\dfrac{ax}{2} + \dfrac{\pi}{4}\right)\right].$

**220.** $\int \sec^2 ax \, dx = \dfrac{1}{a} \tan ax.$

**221.** $\int \sec^3 ax \, dx = \dfrac{1}{2a}\left[\tan ax \sec ax + \log\left(\tan\left(\dfrac{ax}{2} + \dfrac{\pi}{4}\right)\right)\right].$

**222.** $\int \sec^n ax \, dx = \dfrac{1}{a(n-1)} \dfrac{\sin ax}{\cos^{n-1} ax}$

$\qquad\qquad + \dfrac{n-2}{n-1} \int \sec^{n-2} ax \, dx, \quad n \text{ integer} > 1.$

**223.** $\int \csc u \, du = \log(\csc u - \text{ctn } u) = \log\left(\tan \dfrac{u}{2}\right),$

where $u$ is any function of $x$.

**224.** $\int \csc ax \, dx = \dfrac{1}{a} \log\left(\tan \dfrac{ax}{2}\right).$

**225.** $\int \csc^2 ax \, dx = -\dfrac{1}{a} \text{ctn } ax.$

**226.** $\int \csc^3 ax \, dx = \dfrac{1}{2a}\left[-\text{ctn } ax \csc ax + \log\left(\tan\dfrac{ax}{2}\right)\right]$

**227.** $\int \csc^n ax \, dx = -\dfrac{1}{a(n-1)} \dfrac{\cos ax}{\sin^{n-1} ax} + \dfrac{n-2}{n-1} \int \csc^{n-2} ax \, dx, \quad n \text{ integer} > 1.$

## Expressions Containing tan *ax* and sec *ax* or ctn *ax* and csc *ax*

**228.** $\int \tan u \sec u \, du = \sec u, \quad$ where $u$ is any function of $x$.

**229.** $\int \tan ax \sec ax \, dx = \dfrac{1}{a} \sec ax.$

**230.** $\int \tan^n ax \sec^2 ax \, dx = \dfrac{1}{a(n+1)} \tan^{n+1} ax, \quad n \neq -1.$

**231.** $\int \tan\, ax\, \sec^n ax\, dx = \dfrac{1}{an}\sec^n ax, \qquad n \neq 0.$

**232.** $\int \operatorname{ctn} u \csc u\, du = -\csc u, \qquad$ where $u$ is any function of $x$.

**233.** $\int \operatorname{ctn}\, ax \csc\, ax\, dx = -\dfrac{1}{a}\csc\, ax.$

**234.** $\int \operatorname{ctn}^n ax \csc^2 ax\, dx = -\dfrac{1}{a(n+1)}\operatorname{ctn}^{n+1} ax, \qquad n \neq -1.$

**235.** $\int \operatorname{ctn}\, ax \csc^n ax\, dx = -\dfrac{1}{an}\csc^n ax, \qquad n \neq 0.$

**236.** $\int \dfrac{\csc^2 ax\, dx}{\operatorname{ctn}\, ax} = -\dfrac{1}{a}\log(\operatorname{ctn}\, ax).$

# Expressions Containing Algebraic and Trigonometric Functions

**237.** $\int x \sin\, ax\, dx = \dfrac{1}{a^2}\sin\, ax - \dfrac{1}{a}x \cos\, ax.$

**238.** $\int x^2 \sin\, ax\, dx = \dfrac{2x}{a^2}\sin\, ax + \dfrac{2}{a^3}\cos\, ax - \dfrac{x^2}{a}\cos\, ax.$

**239.** $\int x^3 \sin\, ax\, dx = \dfrac{3x^2}{a^2}\sin\, ax - \dfrac{6}{a^4}\sin\, ax - \dfrac{x^3}{a}\cos\, ax + \dfrac{6x}{a^3}\cos\, ax.$

**240.** $\int x \sin^2 ax\, dx = \dfrac{x^2}{4} - \dfrac{x \sin 2ax}{4a} - \dfrac{\cos 2ax}{8a^2}.$

**241.** $\int x^2 \sin^2 ax\, dx = \dfrac{x^3}{6} - \left(\dfrac{x^2}{4a} - \dfrac{1}{8a^3}\right)\sin 2ax - \dfrac{x \cos 2ax}{4a^2}.$

**242.** $\int x^3 \sin^2 ax\, dx = \dfrac{x^4}{8} - \left(\dfrac{x^3}{4a} - \dfrac{3x}{8a^3}\right)\sin 2ax - \left(\dfrac{3x^2}{8a^2} - \dfrac{3}{16a^4}\right)\cos 2ax.$

**243.** $\int x \sin^3 ax\, dx = \dfrac{x \cos 3ax}{12a} - \dfrac{\sin 3ax}{36a^2} - \dfrac{3x \cos ax}{4a} + \dfrac{3 \sin ax}{4a^2}.$

**244.** $\int x^n \sin\, ax\, dx = -\dfrac{1}{a}x^n \cos\, ax + \dfrac{n}{a}\int x^{n-1}\cos\, ax\, dx, \qquad n > 0.$

**245.** $\int \dfrac{\sin\, ax\, dx}{x} = ax - \dfrac{(ax)^3}{3 \cdot 3!} + \dfrac{(ax)^5}{5 \cdot 5!} - \cdots.$

**246.** $\int \dfrac{\sin\, ax\, dx}{x^m} = \dfrac{-1}{(m-1)}\dfrac{\sin\, ax}{x^{m-1}} + \dfrac{a}{(m-1)}\int \dfrac{\cos\, ax\, dx}{x^{m-1}}.$

**247.** $\int x \cos\, ax\, dx = \dfrac{1}{a^2}\cos\, ax + \dfrac{1}{a}x \sin\, ax.$

**248.** $\int x^2 \cos ax \, dx = \frac{2x}{a^2} \cos ax - \frac{2}{a^3} \sin ax + \frac{x^2}{a} \sin ax.$

**249.** $\int x^3 \cos ax \, dx = \frac{(3a^2 x^2 - 6) \cos ax}{a^4} + \frac{(a^2 x^3 - 6x) \sin a}{a^3}.$

**250.** $\int x \cos^2 ax \, dx = \frac{x^2}{4} + \frac{x \sin 2ax}{4a} + \frac{\cos 2ax}{8a^2}.$

**251.** $\int x^2 \cos^2 ax \, dx = \frac{x^3}{6} + \left( \frac{x^2}{4a} - \frac{1}{8a^3} \right) \sin 2ax + \frac{x \cos 2ax}{4a^2}.$

**252.** $\int x^3 \cos^2 ax \, dx = \frac{x^4}{8} + \left( \frac{x^3}{4a} - \frac{3x}{8a^3} \right) \sin 2ax + \left( \frac{3x^2}{8a^2} - \frac{3}{16a^4} \right) \cos 2ax.$

**253.** $\int x \cos^3 ax \, dx = \frac{x \sin 3ax}{12a} + \frac{\cos 3ax}{36a^2} + \frac{3x \sin ax}{4a} + \frac{3 \cos ax}{4a^2}.$

**254.** $\int x^n \cos ax \, dx = \frac{1}{a} x^n \sin ax - \frac{n}{a} \int x^{n-1} \sin ax \, dx, \quad n \text{ pos.}$

**255.** $\int \frac{\cos ax \, dx}{x} = \log(ax) - \frac{(ax)^2}{2 \cdot 2!} + \frac{(ax)^4}{4 \cdot 4!} - \cdots.$

**256.** $\int \frac{\cos ax}{x^m} \, dx = -\frac{1}{(m-1)} \frac{\cos ax}{x^{m-1}} - \frac{a}{(m-1)} \int \frac{\sin ax \, dx}{x^{m-1}}.$

---

# Expressions Containing Exponential and Logarithmic Functions

**257.** $\int e^u \, du = e^u, \quad$ where $u$ is any function of $x.$

**258.** $\int b^u \, du = \frac{b^u}{\log(b)}, \quad$ where $u$ is any function of $x.$

**259.** $\int e^{ax} \, dx = \frac{1}{a} e^{ax}, \qquad \int b^{ax} \, dx = \frac{b^{ax}}{a \log(b)}.$

**260.** $\int xe^{ax} \, dx = \frac{e^{ax}}{a^2} (ax - 1), \qquad \int xb^{ax} \, dx = \frac{xb^{ax}}{a \log(b)} - \frac{b^{ax}}{a^2 (\log(b))^2}.$

**261.** $\int x^2 e^{ax} \, dx = \frac{e^{ax}}{a^3} (a^2 x^2 - 2ax + 2).$

**262.** $\int x^n e^{ax} \, dx = \frac{1}{a} x^n e^{ax} - \frac{n}{a} \int x^{n-1} e^{ax} \, dx, \quad n \text{ pos.}$

**263.** $\int x^n e^{ax} \, dx = \frac{e^{ax}}{a^{n+1}} [(ax)^n - n(ax)^{n-1} + n(n-1)(ax)^{n-2}$
$$- \cdots + (-1)^n n!], \quad n \text{ pos. integ.}$$

**264.** $\int x^n e^{-ax}\, dx = -\dfrac{e^{-ax}}{a^{n+1}}[(ax)^n + n(ax)^{n-1} + n(n-1)(ax)^{n-2}$
$$+ \cdots + n!], \quad n \text{ pos. integ.}$$

**265.** $\int x^n b^{ax}\, dx = \dfrac{x^n b^{ax}}{a \log(b)} - \dfrac{n}{a \log(b)} \int x^{n-1} b^{ax}\, dx, \quad n \text{ pos.}$

**266.** $\int \dfrac{e^{ax}}{x}\, dx = \log(x) + ax + \dfrac{(ax)^2}{2 \cdot 2!} + \dfrac{(ax)^3}{3 \cdot 3!} + \cdots.$

**267.** $\int \dfrac{e^{ax}}{x^n}\, dx = \dfrac{1}{n-1}\left(-\dfrac{e^{ax}}{x^{n-1}} + a \int \dfrac{e^{ax}}{x^{n-1}}\, dx\right), \quad n \text{ integ.} > 1.$

**268.** $\int \dfrac{dx}{b + ce^{ax}} = \dfrac{1}{ab}[ax - \log(b + ce^{ax})].$

**269.** $\int \dfrac{e^{ax}\, dx}{b + ce^{ax}} = \dfrac{1}{ac} \log(b + ce^{ax}).$

**270.** $\int \dfrac{dx}{be^{ax} + ce^{-ax}} = \dfrac{1}{a\sqrt{bc}} \tan^{-1}\left(e^{ax}\sqrt{\dfrac{b}{c}}\right), \quad b \text{ and } c \text{ pos.}$

**271.** $\int e^{ax} \sin bx\, dx = \dfrac{e^{ax}}{a^2 + b^2}(a \sin bx - b \cos bx).$

**272.** $\int e^{ax} \sin bx \sin cx\, dx = \dfrac{e^{ax}[(b-c)\sin(b-c)x + a \cos(b-c)x]}{2[a^2 + (b-c)^2]}$
$$- \dfrac{e^{ax}[(b+c)\sin(b+c) + a\cos(b+c)x]}{2[a^2 + (b+c)^2]}.$$

**273.** $\int e^{ax} \cos bx\, dx = \dfrac{e^{ax}}{a^2 + b^2}(a \cos bx + b \sin bx).$

**274.** $\int e^{ax} \cos bx \cos cx\, dx = \dfrac{e^{ax}[(b-c)\sin(b-c)x + a \cos(b-c)x]}{2[a^2 + (b-c)^2]}$
$$+ \dfrac{e^{ax}[(b+c)\sin(b+c)x + a\cos(b+c)x]}{2[a^2 + (b+c)^2]}.$$

**275.** $\int e^{ax} \sin bx \cos cx\, dx = \dfrac{e^{ax}[a\sin(b-c)x - (b-c)\cos(b-c)x]}{2[a^2 + (b-c)^2]}$
$$+ \dfrac{e^{ax}[a\sin(b+c)x - (b+c)\cos(b+c)x]}{2[a^2 + (b+c)^2]}.$$

**276.** $\int e^{ax} \sin bx \sin(bx + c)\, dx = \dfrac{e^{ax} \cos c}{2a}$
$$- \dfrac{e^{ax}[a\cos(2bx + c) + 2b\sin(2bx + c)]}{2(a^2 + 4b^2)}.$$

**277.** $\int e^{ax} \cos bx \cos(bx + c)\, dx = \dfrac{e^{ax} \cos c}{2a}$
$$+ \dfrac{e^{ax}[a\cos(2bx + c) + 2b\sin(2bx + c)]}{2(a^2 + 4b^2)}.$$

**278.** $\int e^{ax} \sin bx \cos(bx + c) \; dx = -\dfrac{e^{ax} \sin c}{2a}$

$$+ \dfrac{e^{ax}[a \sin(2bx + c) - 2b \cos(2bx + c)]}{2(a^2 + 4b^2)}.$$

**279.** $\int e^{ax} \cos bx \sin(bx + c) \; dx = \dfrac{e^{ax} \sin c}{2a}$

$$+ \dfrac{e^{ax}[a \sin(2bx + c) - 2b \cos(2bx + c)]}{2(a^2 + 4b^2)}.$$

**280.** $\int xe^{ax} \sin bx \; dx = \dfrac{xe^{ax}}{a^2 + b^2}(a \sin bx - b \cos bx)$

$$- \dfrac{e^{ax}}{(a^2 + b^2)^2}[(a^2 - b^2) \sin bx - 2\,ab \cos bx].$$

**281.** $\int xe^{ax} \cos bx \; dx = \dfrac{xe^{ax}}{a^2 + b^2}(a \cos bx + b \sin bx)$

$$- \dfrac{e^{ax}}{(a^2 + b^2)^2}[(a^2 - b^2) \cos bx + 2\,ab \sin bx].$$

**282.** $\int e^{ax} \cos^n bx \; dx = \dfrac{e^{ax} (\cos^{n-1} bx)(a \cos bx + nb \sin bx)}{a^2 + n^2 b^2}$

$$+ \dfrac{n(n - 1)b^2}{a^2 + n^2 b^2} \int e^{ax} \cos^{n-2} bx \; dx.$$

**283.** $\int e^{ax} \sin^n bx \; dx = \dfrac{e^{ax} (\sin^{n-1} bx)(a \sin bx - nb \cos bx)}{a^2 + n^2 b^2}$

$$+ \dfrac{n(n - 1)b^2}{a^2 + n^2 b^2} \int e^{ax} \sin^{n-2} bx \; dx.$$

**284.** $\int \log(ax) \; dx = x \log(ax) - x.$

**285.** $\int x \log(ax) \; dx = \dfrac{x^2}{2} \log(ax) - \dfrac{x^2}{4}.$

**286.** $\int x^2 \log(ax) \; dx = \dfrac{x^3}{3} \log(ax) - \dfrac{x^3}{9}.$

**287.** $\int [\log(ax)]^2 \; dx = x[\log(ax)]^2 - 2x \log(ax) + 2x.$

**288.** $\int [\log(ax)]^n \; dx = x[\log(ax)]^n - n \int [\log(ax)]^{n-1} \; dx, \quad n \text{ pos.}$

**289.** $\int x^n \log(ax) \; dx = x^{n+1}\left(\dfrac{\log(ax)}{n + 1} - \dfrac{1}{(n + 1)^2}\right), \quad n \neq -1.$

**290.** $\int x^n [\log(ax)]^m \; dx = \dfrac{x^{n+1}}{n + 1}[\log(ax)]^m - \dfrac{m}{n + 1} \int x^n [\log(ax)]^{m-1} \; dx.$

**291.** $\int \dfrac{[\log(ax)]^n}{x}\, dx = \dfrac{[\log(ax)]^{n+1}}{n+1}, \qquad n \neq -1.$

**292.** $\int \dfrac{dx}{x\log(ax)} = \log[\log(ax)].$

**293.** $\int \dfrac{dx}{x(\log(ax))^n} = -\dfrac{1}{(n-1)(\log(ax))^{n-1}}.$

**294.** $\int \dfrac{x^n\, dx}{(\log(ax))^m} =$

$$\dfrac{-x^{n+1}}{(m-1)(\log(ax))^{m-1}} + \dfrac{n+1}{m-1}\int \dfrac{x^n\, dx}{(\log(ax))^{m-1}}, \qquad m \neq 1.$$

**295.** $\int \dfrac{x^n\, dx}{\log(ax)} = \dfrac{1}{a^{n+1}}\int \dfrac{e^y\, dy}{y}, \qquad y = (n+1)\ln|ax|.$

**296.** $\int \dfrac{x^n\, dx}{\log(ax)} = \dfrac{1}{a^{n+1}}\big[\log\big|\log(ax)\big| + (n+1)\log(ax)$

$$+ \dfrac{(n+1)^2[\log(ax)]^2}{2\cdot 2!} + \dfrac{(n+1)^3[\log(ax)]^3}{3\cdot 3!} + \cdots\big].$$

**297.** $\int \dfrac{dx}{\log(ax)} =$

$$\dfrac{1}{a}\big[\log[\log(ax)] + \log(ax) + \dfrac{(\log(ax))^2}{2\cdot 2!} + \dfrac{(\log(ax))^3}{3\cdot 3!} + \cdots\big].$$

**298.** $\int \sin[\log(ax)]\, dx = \dfrac{x}{2}[\sin[\log(ax)] - \cos[\log(ax)]].$

**299.** $\int \cos[\log(ax)]\, dx = \dfrac{x}{2}[\sin[\log(ax)] + \cos[\log(ax)]].$

**300.** $\int e^{ax}\log(bx)\, dx = \dfrac{1}{a}e^{ax}\log(bx) - \dfrac{1}{a}\int \dfrac{e^{ax}}{x}\, dx.$

## Expressions Containing Inverse Trigonometric Functions

**301.** $\int \sin^{-1}ax\, dx = x\sin^{-1}ax + \dfrac{1}{a}\sqrt{1 - a^2x^2}.$

**302.** $\int (\sin^{-1}ax)^2\, dx = x(\sin^{-1}ax)^2 - 2x + \dfrac{2}{a}\sqrt{1 - a^2x^2}\,\sin^{-1}ax.$

**303.** $\int x\sin^{-1}ax\, dx = \dfrac{x^2}{2}\sin^{-1}ax - \dfrac{1}{4a^2}\sin^{-1}ax + \dfrac{x}{4a}\sqrt{1 - a^2x^2}.$

**304.** $\int x^n\sin^{-1}ax\, dx = \dfrac{x^{n+1}}{n+1}\sin^{-1}ax - \dfrac{a}{n+1}\int \dfrac{x^{n+1}\, dx}{\sqrt{1 - a^2x^2}}, \qquad n \neq -1.$

**305.** $\displaystyle\int \frac{\sin^{-1}ax\, dx}{x} = ax + \frac{1}{2\cdot3\cdot3}(ax)^3 + \frac{1\cdot3}{2\cdot4\cdot5\cdot5}(ax)^5$
$$+ \frac{1\cdot3\cdot5}{2\cdot4\cdot6\cdot7\cdot7}(ax)^7 + \cdots, \qquad a^2x^2 < 1.$$

**306.** $\displaystyle\int \frac{\sin^{-1}ax\, dx}{x^2} = -\frac{1}{x}\sin^{-1}ax - a\log\!\left(\frac{1 + \sqrt{1 - a^2x^2}}{ax}\right).$

**307.** $\displaystyle\int \cos^{-1}ax\, dx = x\cos^{-1}ax - \frac{1}{a}\sqrt{1 - a^2x^2}.$

**308.** $\displaystyle\int (\cos^{-1}ax)^2\, dx = x(\cos^{-1}ax)^2 - 2x - \frac{2}{a}\sqrt{1 - a^2x^2}\,\cos^{-1}ax.$

**309.** $\displaystyle\int x\cos^{-1}ax\, dx = \frac{x^2}{2}\cos^{-1}ax - \frac{1}{4a^2}\cos^{-1}ax - \frac{x}{4a}\sqrt{1 - a^2x^2}.$

**310.** $\displaystyle\int x^n\cos^{-1}ax\, dx = \frac{x^{n+1}}{n+1}\cos^{-1}ax + \frac{a}{n+1}\int \frac{x^{n+1}\, dx}{\sqrt{1 - a^2x^2}}, \qquad n \neq -1.$

**311.** $\displaystyle\int \frac{\cos^{-1}ax\, dx}{x} = \frac{\pi}{2}\ln|ax| - ax - \frac{1}{2\cdot3\cdot3}(ax)^3$
$$- \frac{1\cdot3}{2\cdot4\cdot5\cdot5}(ax)^5 - \frac{1\cdot3\cdot5}{2\cdot4\cdot6\cdot7\cdot7}(ax)^7 - \cdots, \qquad a^2x^2 < 1.$$

**312.** $\displaystyle\int \frac{\cos^{-1}ax\, dx}{x^2} = -\frac{1}{x}\cos^{-1}ax + a\log\!\left(\frac{1 + \sqrt{1 - a^2x^2}}{ax}\right).$

**313.** $\displaystyle\int \tan^{-1}ax\, dx = x\tan^{-1}ax - \frac{1}{2a}\log(1 + a^2x^2).$

**314.** $\displaystyle\int x^n\tan^{-1}ax\, dx = \frac{x^{n+1}}{n+1}\tan^{-1}ax - \frac{a}{n+1}\int \frac{x^{n+1}\, dx}{1 + a^2x^2}, \qquad n \neq -1.$

**315.** $\displaystyle\int \frac{\tan^{-1}ax\, dx}{x^2} = -\frac{1}{x}\tan^{-1}ax - \frac{a}{2}\log\!\left(\frac{1 + a^2x^2}{a^2x^2}\right).$

**316.** $\displaystyle\int \mathrm{ctn}^{-1}ax\, dx = x\,\mathrm{ctn}^{-1}ax + \frac{1}{2a}\log(1 + a^2x^2).$

**317.** $\displaystyle\int x^n\,\mathrm{ctn}^{-1}ax\, dx = \frac{x^{n+1}}{n+1}\mathrm{ctn}^{-1}ax + \frac{a}{n+1}\int \frac{x^{n+1}\, dx}{1 + a^2x^2}, \qquad n \neq -1.$

**318.** $\displaystyle\int \frac{\mathrm{ctn}^{-1}ax\, dx}{x^2} = -\frac{1}{x}\mathrm{ctn}^{-1}ax + \frac{a}{2}\log\!\left(\frac{1 + a^2x^2}{a^2x^2}\right).$

**319.** $\displaystyle\int \sec^{-1}ax\, dx = x\sec^{-1}ax - \frac{1}{a}\log(ax + \sqrt{a^2x^2 - 1}).$

**320.** $\displaystyle\int x^n\sec^{-1}ax\, dx = \frac{x^{n+1}}{n+1}\sec^{-1}ax \pm \frac{1}{n+1}\int \frac{x^n\, dx}{\sqrt{a^2x^2 - 1}}, \qquad n \neq -1;$

use + sign when $\pi/2 < \sec^{-1}ax < \pi$; − sign when $0 < \sec^{-1}ax < \pi/2$.

**321.** $\int \csc^{-1} ax \, dx = x \csc^{-1} ax + \dfrac{1}{a} \log(ax + \sqrt{a^2 x^2 - 1}).$

**322.** $\int x^n \csc^{-1} ax \, dx = \dfrac{x^{n+1}}{n+1} \csc^{-1} ax \pm \dfrac{1}{n+1} \int \dfrac{x^n \, dx}{\sqrt{a^2 x^2 - 1}}, \qquad n \neq -1;$

use $+$ sign when $0 < \csc^{-1} ax < \pi/2$; $-$ sign when $-\pi/2 < \csc^{-1} ax < 0.$

# *Index*